符号模式矩阵允许对角化问题研究进展

The Recent Research of Sign Pattern That Allows Diagonalizability

冯新磊 • 著

图书在版编目（CIP）数据

符号模式矩阵允许对角化问题研究进展 ：英文 / 冯新磊著. — 成都 ：四川大学出版社，2023.10
ISBN 978-7-5690-6322-6

Ⅰ. ①符… Ⅱ. ①冯… Ⅲ. ①对角化－英文 Ⅳ. ①O241.6

中国国家版本馆 CIP 数据核字（2023）第 162834 号

书　　名：符号模式矩阵允许对角化问题研究进展
Fuhao Moshi Juzhen Yunxu Duijiaohua Wenti Yanjiu Jinzhan
著　　者：冯新磊

选题策划：孙明丽　梁　平
责任编辑：孙明丽
责任校对：唐　飞
装帧设计：裴菊红
责任印制：王　炜

出版发行：四川大学出版社有限责任公司
地址：成都市一环路南一段 24 号（610065）
电话：（028）85408311（发行部）、85400276（总编室）
电子邮箱：scupress@vip.163.com
网址：https://press.scu.edu.cn
印前制作：四川胜翔数码印务设计有限公司
印刷装订：四川省平轩印务有限公司

成品尺寸：170mm×240mm
印　　张：7.25
字　　数：179 千字

版　　次：2023 年 11 月 第 1 版
印　　次：2023 年 11 月 第 1 次印刷
定　　价：46.00 元

扫码获取数字资源

四川大学出版社
微信公众号

Preface

Characterization of sign patterns that allow diagonalizability has been a long-standing open problem. In this book, we mainly considered this problem, and obtained some sufficient and/or necessary conditions for a sign pattern to allow diagonalizability. Meanwhile, we also introduced the research history of this problem and other researchers' results.

In the following we will introduce the content of this book.

In Chapter 1, we mainly introduced the research history of sign patterns that allow diagonalizability, and preliminaries knowledge. Moreover, we also briefly give out the other researcher's important results. In Chapter 2 – 6, we presented our results of sign patterns that allow diagonalizability except Section 6. 3.

In Chapter 2, in order to obtain these conditions, we showed that a 2×2 complex matrix $\boldsymbol{A}$ is diagonally equivalent to a matrix with two distinct eigenvalues if and only if $\boldsymbol{A}$ is not strictly triangular. It is established that every 3×3 non-singular matrix is diagonally equivalent to a matrix with 3 distinct eigenvalues. More precisely, a 3×3 matrix $\boldsymbol{A}$ is not diagonally equivalent to any matrix with 3 distinct eigenvalues if and only if $\det \boldsymbol{A} = 0$ and each principal minor of $\boldsymbol{A}$ of order 2 is zero. It is conjectured that for all $n \geqslant 2$, an $n \times n$ complex matrix is not diagonally equivalent to any matrix with n distinct eigenvalues if and only if $\det \boldsymbol{A} = 0$ and every principal minor of $\boldsymbol{A}$ of order $n-1$ is zero. Thus some necessary and sufficient conditions for a sign pattern to allow diagonalizability are obtained, in terms of allowing related properties. Some properties of normal sign patterns are considered. In particular, it is shown that normal sign patterns of order up to 3 allow diagonalizability. Two combinatorial necessary conditions for a sign pattern to allow diagonalizability are also presented.

In Chapter 3, we also obtained some results between minimum or maximum rank allowing diagonalizability. It is also known that for each $k \geqslant 4$, there exists an irreducible sign pattern with minimum rank k that does not allow diagonalizability. However, it is shown in this chapter that every square sign pattern $\boldsymbol{A}$ with minimum rank 2 that has no zero line allows diagonalizability with rank 2 and also with rank equal to the maximum rank of the sign pattern. In particular, every irreducible sign pattern with minimum rank 2 allows diagonalizability. On the other hand, an example is given to show the existence of a square sign pattern with minimum rank 3 and no zero line that does not allow diagonalizability; however, the case for irreducible sign patterns with minimum rank 3 remains open. In addition, for a sign pattern that allows diagonalizability, the possible ranks of the diagonalizable real matrices with the specified sign pattern are shown to be lengths of certain composite cycles. Some results on sign patterns with minimum rank 2 are extended to sign pattern matrices whose maximal zero submatrices are strongly disjoint (that is, their row index sets as well as their column index sets are pairwise disjoint).

In Chapter 4, we obtained some new necessary and sufficient conditions which a sign k-potent sign pattern matrix allows sign k-potent, and the result which a sign k-potent sign pattern matrix that allows sign k-potent also allows diagonalizability.

In Chapter 5, we further considered how many entries need to be changed to obtain a matrix $\boldsymbol{B}_0 \in \mathrm{Q}(\boldsymbol{A})$ with rank $MR(\boldsymbol{A})$ from a matrix $\boldsymbol{B} \in Q(\boldsymbol{A})$ with rank $mr(\boldsymbol{A})$ or contrary case, and obtained some new results about sign pattern that allows diagonalizability. Finally, we also obtained some results on a sign pattern matrix in Frobenius normal form that allows diagonalizability.

In Chapter 6, the necessary and sufficient conditions that sign patterns allow diagonalizability from combinational opinion are still an open problem. Therefore, we made other efforts about this problem. Firstly, we made a carefully discussion about allowing unitary diagonalizability of two sign pattern. Some sufficient and necessary conditions of allowing unitary diagonalizability are also obtained. Second, we obtained some results of 1 - 4 order sign patterns that allow diagonalizability. Through this way we want to find its rules and finally to solve this problem. However, we find that it is difficult for higher order sign patterns. Finally we also introduced some latest researches of this problem.

Content

Chapter 1

Research History of Sign Patterns That Allow Diagonalizability

1.1 Introduction

The origins of sign pattern matrices are the need to solve certain problems in economics and other areas based only on the signs of the entries of the matrices in the book by the Nobel Economics Prize winner P. Samuelson[36]. Sign pattern matrices have been heavily studied and have found applications in other areas (see [4, 10, 21]). In particular, various eigenvalue problems played important roles in both traditional matrix theory and sign pattern matrix theory[21,23].

From 1947, sign pattern mainly has several research directions, such as sign-solvability, system stability, sign pattern allowing or requiring properties, sign pattern minimum rank, inertia, complex sign patterns, ray patterns, powers of sign patterns, where sign pattern allowing properties includes diagonalizability, certain types of inverses, nilpotence, normality, orthogonality, et al.

In this book, the researcher mainly considers the problems of sign patterns that allow diagonalizability. In the following we will introduce the research history of sign patterns that allows diagonalizability.

It is well known that a matrix all of whose eigenvalues are distinct is diagonalizable[24]. The sign patterns that require all eigenvalues to be distinct have been studied by Li and Harris[30]. The search for sufficient and necessary conditions characterizing sign patterns that allow diagonalizability has been a long-standing open problem, studied by Eschenbach and Johnson in [10], Shao and Gao in [37], Feng and Li et al. in [13 - 17], Das in [7].

In 1993, Eschenbach and Johnson[10] considered this problem, and obtained the following results.

Theorem 1.1.1[10] Let $\boldsymbol{A}$ be an $n \times n$ sign pattern. If the maximum composite cycle length in $\boldsymbol{A}$ is equal to the maximum rank of $\boldsymbol{A}$, then $\boldsymbol{A}$ allows diagonalizability. In particular, if the maximum composite cycle length of $\boldsymbol{A}$ is at least $n-1$, then $\boldsymbol{A}$ allows diagonalizability.

Moreover, they also presented a necessary condition in [10].

Theorem 1.1.2[10] If a sign pattern $\boldsymbol{A}$ allows diagonalizability, then the minimum rank of $\boldsymbol{A}$ is less than or equal to the maximum composite cycle length of $\boldsymbol{A}$.

However, this necessary condition is not sufficient. Moreover, if a sign pattern $\boldsymbol{A}$ does not satisfy conditions of Theorem 1.1.2, then we can introduce a suitable permutation $\boldsymbol{P}$ such that $\boldsymbol{PA}$ satisfies the condition of Theorem 1.1.2. Therefore, we can get the following results.

Theorem 1.1.3[10] For any $n \times n$ sign pattern matrix $\boldsymbol{A}$, there is a permutation matrix $\boldsymbol{P}$ such that $\boldsymbol{PA}$ allows diagonalizability.

In [10], Eschenbach and Johnson have presented that $z(\boldsymbol{A}) \leqslant D(\boldsymbol{A})$ is also a necessary condition for a sign pattern that allowing diagonalizability, and thought that this necessary condition is also sufficient. Meanwhile, Shao and Gao[36] presented two counterexamples and showed that this conjecture are wrong in 2003.

For example, let

$$\boldsymbol{A} = \begin{bmatrix} 0 & + & - & 0 & 0 \\ + & 0 & + & 0 & 0 \\ + & + & 0 & 0 & 0 \\ 0 & 0 & 0 & 0 & + \\ 0 & 0 & 0 & 0 & 0 \end{bmatrix}.$$

We can get $z(\boldsymbol{A}) = D(\boldsymbol{A})$, and $\boldsymbol{A}$ does not allow diagonalizability.

Furthermore, in 2006 Shao and Gao[38] gave in detail that the conjectrue is wrong. Their results are the following:

(1) For a $n \times n$ reducible sign patterns, Eschenbach - Johnson Conjectrue is right for $n \leqslant 3$, and not true for $n \geqslant 4$;

(2) For a $n \times n$ irreducible sign patterns, Eschenbach - Johnson Conjectrue is right for $n \leqslant 5$, and not true for $n \geqslant 6$.

From these results, we also saw that an irreducible sign pattern may not be allowed diagonalizability. Moreover, for irreducible sign patterns they also got the following results:

Theorem 1.1.4[38] Let $\boldsymbol{A}$ be an $n \times n$ irreducible sign patterns, and $2 \leqslant n \leqslant 4$, then $\boldsymbol{A}$ allows diagonalizability.

Theorem 1.1.5[38] Let $\boldsymbol{A}$ be an $n \times n$ irreducible sign patterns with $z(\boldsymbol{A}) \leqslant D(\boldsymbol{A})$, and $n = 5$, then $\boldsymbol{A}$ allows diagonalizability.

In 2002, Li and Harries[30] considered the problem that sign pattern matrices require all distinct eigenvalues. We know that matrices with all distinct eigenvalues can be diagonalizable. They gave the following results:

Theorem 1.1.6 [30] If A is an $n \times n$ symmetric irreducible tridiagonal sign pattern matrix, then A requires all distinct real eigenvalues.

Theorem 1.1.7 [30] If A is an $n \times n$ skew symmetric irreducible tridiagonal sign pattern matrix, then A requires all distinct pure imaginary (possibly including zero) eigenvalues.

Theorem 1.1.8 [30] If A is an 2×2 irreducible sign pattern requires two distinct eigenvalues if and only if it is symmetric or skew-symmetric.

They also got the results that all the 3×3 irreducible sign patterns that require distinct eigenvalues, which are the irreducible tridiagonal symmetric sign patterns, irreducible tridiagonal skew-symmetric sign patterns, and 3-cycle sign patterns with the following forms:

$$\begin{bmatrix} + & + & 0 \\ 0 & 0 & + \\ + & 0 & 0 \end{bmatrix}, \begin{bmatrix} 0 & + & 0 \\ - & 0 & + \\ + & - & 0 \end{bmatrix}, \begin{bmatrix} 0 & + & 0 \\ - & 0 & + \\ + & 0 & 0 \end{bmatrix}, \begin{bmatrix} 0 & + & - \\ - & 0 & + \\ + & - & 0 \end{bmatrix}, \begin{bmatrix} + & + & 0 \\ 0 & 0 & + \\ + & - & 0 \end{bmatrix}.$$

In 2011, Kim further considered this problem, and obtained the following results:

Theorem 1.1.9 [27] If A is an 4×4 irreducible sign pattern requires four distinct real eigenvalues, then up to equivalence, A is the following forms:

(1) Symmetric tridiagonal sign patterns

$$\begin{bmatrix} * & + & 0 & 0 \\ + & * & + & 0 \\ 0 & + & * & + \\ 0 & 0 & + & * \end{bmatrix},$$

where * maybe +, − or 0.

(2) Symmetric star sign patterns

$$\begin{bmatrix} a_1 & + & + & + \\ + & a_2 & 0 & 0 \\ + & 0 & a_3 & 0 \\ + & 0 & 0 & a_4 \end{bmatrix},$$

where a_i, $i = 1, 2, 3, 4$, maybe +, − or 0, and a_2, a_3 and a_4 are not the same.

Theorem 1.1.10[27] If $\boldsymbol{A}$ is an 4×4 irreducible sign pattern requires four distinct non-real eigenvalues, then up to equivalence, $\boldsymbol{A}$ is the following forms:

$$\begin{bmatrix} 0 & + & 0 & 0 \\ - & 0 & + & 0 \\ 0 & - & 0 & + \\ 0 & 0 & - & 0 \end{bmatrix} \text{ or } \begin{bmatrix} 0 & + & 0 & a_1 \\ -a_2 & 0 & + & 0 \\ 0 & -a_3 & 0 & + \\ - & 0 & -a_4 & 0 \end{bmatrix},$$

where a_i, $i=1, 2, 3, 4$, maybe + or 0.

Moreover, Shao and Gao[37] also got the result that if $\boldsymbol{A}$ is combinatorially symmetric, then it allows diagonalizability. This results will be introduced in the following sections.

In 2012, in order to obtain these results, we conjectured that for all $n\geqslant 2$, an $n\times n$ complex matrix is not diagonally equivalent to any matrix with n distinct eigenvalues if and only if $\det \boldsymbol{A}=0$ and every principal minor of $\boldsymbol{A}$ of order $n-1$ is zero[13]. In 2013, some necessary and sufficient conditions for a sign pattern that allow diagonalizability are obtained[14], based on the Conjecture of [13]. Some properties of normal sign patterns are considered. In particular, it is shown that normal sign patterns of order up to 3 allow diagonalizability. Two combinatorial necessary conditions for a sign pattern to allow diagonalizability are also presented.

In 2017, we obtained some necessary and sufficient conditions which a sign k-potent sign pattern matrix allows sign k-potent, so we got that a sign k-potent sign pattern matrix allows sign k-potent also allows diagonalizability[15].

In 2020, we also obtained some results between minimum (or maximum rank) and allowing diagonalizability[16]. It is shown in this paper that every square sign pattern $\boldsymbol{A}$ with minimum rank 2 that has no zero line allows diagonalizability with rank 2 and also with rank equal to the maximum rank of the sign pattern. In particular, every irreducible sign pattern with minimum rank 2 allows diagonalizability. On the other hand, an example is given to show the existence of a square sign pattern with minimum rank 3 and no zero line that does not allow diagonalizability. In addition, for a sign pattern that allows diagonalizability, the possible ranks of the diagonalizable real matrices with the specified sign pattern are shown to be lengths of certain composite cycles. Some results on sign patterns with minimum rank 2 are extended to sign pattern matrices whose maximal zero sub-matrices are strongly disjoint.

In 2022, we further considered how many entries need to be changed to obtain a matrix $\boldsymbol{B}_0 \in Q(\boldsymbol{A})$ with rank $MR(\boldsymbol{A})$ from a matrix $\boldsymbol{B} \in Q(\boldsymbol{A})$ with rank $mr(\boldsymbol{A})$ or contrary case. And obtained some new results of sign pattern that allow diagonalizability[17]. Finally, we also obtained some results on a sign pattern matrix in Frobenius normal form that allows diagonalizability. Moreover, Das[7] further considered sign patterns that allow diagonalizability whose graphs are star or path, and presented a sufficient condition of sign pattern matrices that allow diagonalizability whose graphs are trees.

Moreover, we also made other efforts about this problems from different research direction. First, we made a careful discussion about allowing unitary diagonalizability of two sign pattern. Some sufficient and necessary conditions of allowing unitary diagonalizability are also obtained. Second, we obtained some results of 1 − 4 order sign patterns that allow diagonalizability. Through this way we want to find its laws and finally to solve this problem. Finally, we find that this way is also difficult. But we will continue to do these works in the late.

1.2 Preliminaries

We now introduce some definitions and notation, most of which can be found in [9, 10, 21, 37].

A sign pattern (matrix) is a matrix whose entries are from the set $\{+, -, 0\}$. The set of all $n \times n$ sign patterns is denoted by Q_n. For $\boldsymbol{A} = [a_{ij}] \in Q_n$, associated with $\boldsymbol{A}$ is a class of real matrices, called the qualitative class of $\boldsymbol{A}$, defined by $Q(\boldsymbol{A}) = \{\boldsymbol{B} = [b_{ij}] \in M_n(R) \mid \operatorname{sgn} b_{ij} = a_{ij} \text{ for all } i \text{ and } j\}$. We may indicate the fact that $\boldsymbol{B} \in Q(\boldsymbol{A})$ by writing $\operatorname{sgn}(\boldsymbol{B}) = \boldsymbol{A}$.

A sign pattern $\boldsymbol{A} = [a_{ij}] \in Q_n$ is said to be combinatorially symmetric if $a_{ij} \neq 0$ iff $a_{ji} \neq 0$, for all i, j.

Ageneralized sign pattern (matrix) is a matrix whose entries are in the set $\{+, -, 0, \#\}$, where # indicates an ambiguous sum (the result of adding + with −). The concept of qualitative class (as well as many other terms mentioned in this section) may be extended to generalized sign patterns, by allowing the entries corresponding to the # positions to take arbitrary real values. In this book we mainly consider sign patterns. A product of sign patterns may be a generalized sign pattern. For example, $\begin{bmatrix} + & + \\ - & + \end{bmatrix}^2 = \begin{bmatrix} \# & + \\ - & \# \end{bmatrix}$.

We say that two $n \times 1$ sign vectors $\boldsymbol{\mu}$ and $\boldsymbol{v}$ match if $\boldsymbol{\mu}^{\mathrm{T}}\boldsymbol{v}$ is non-zero and unambiguously signed, that is, $\boldsymbol{\mu}^{\mathrm{T}}\boldsymbol{v}$ is + or −. If $\boldsymbol{\mu}^{\mathrm{T}}\boldsymbol{v} = +$ (respectively, $\boldsymbol{\mu}^{\mathrm{T}}\boldsymbol{v} = -$) we say that $\boldsymbol{\mu}$ and $\boldsymbol{v}$ are positively matched (respectively, negatively matched).

Let $\boldsymbol{P}$ be a property referring to a real matrix. For a sign pattern $\boldsymbol{A}$, if there exists a real matrix $\boldsymbol{B} \in Q(\boldsymbol{A})$ such that $\boldsymbol{B}$ has property $\boldsymbol{P}$, we say that $\boldsymbol{A}$ allows $\boldsymbol{P}$; if every $\boldsymbol{B} \in Q(\boldsymbol{A})$ has property $\boldsymbol{P}$, we say that $\boldsymbol{A}$ requires $\boldsymbol{P}$.

The signed digraph of an $n \times n$ sign pattern $\boldsymbol{A} = [a_{ij}]$, denoted by $D(\boldsymbol{A})$, is the digraph with vertex set $\{1, 2, \cdots, n\}$, where (i, j) is an arc if only and if $a_{ij} \neq 0$. Let $\boldsymbol{A} = [a_{ij}]$ be an $n \times n$ sign pattern. A non-zero product of the form $\gamma = a_{i_1 i_2} a_{i_2 i_3} \cdots a_{i_k i_1}$, in which the indices $i_1, i_2, \cdots, i_k$ are distinct, is called a simple cycle of length k (or a k-cycle). Each $i_m (m = 1, 2, \cdots, k)$ is called a vertex of γ. A composite cycle of length k is a product of simple cycles whose total length is k and whose index sets are mutually disjoint.

The largest possible length of the composite cycles of $\boldsymbol{A}$ is called the maximum cycle length of $\boldsymbol{A}$, denoted by$c(\boldsymbol{A})$. If $\boldsymbol{A}$ has no simple cycle at all, then $c(\boldsymbol{A}) = 0$.

Let $\boldsymbol{A} \in Q_n$. We define $MR(\boldsymbol{A})$, the maximal rank of A by

$$MR(\boldsymbol{A}) = \max\{\operatorname{rank} \boldsymbol{B} \mid \boldsymbol{B} \in Q(\boldsymbol{A})\}.$$

Similarly, the minimal rank of $\boldsymbol{A}$, $mr(\boldsymbol{A})$, is given by

$$mr(\boldsymbol{A}) = \min\{\operatorname{rank} \boldsymbol{B} \mid \boldsymbol{B} \in Q(\boldsymbol{A})\}.$$

A permutation sign pattern is a square sign pattern matrix with entries 0 and +, where the entry + occurs precisely once in each row and in each column. Note that a permutation sign pattern $\boldsymbol{P} \in Q_n$ satisfies $\boldsymbol{P}^{\mathrm{T}}\boldsymbol{P} = \boldsymbol{P}\boldsymbol{P}^{\mathrm{T}} = \boldsymbol{I}_n$, where $\boldsymbol{I}_n$ is the identity sign pattern of order n, namely, $\boldsymbol{I}_n$ is the diagonal sign pattern of order n all of whose diagonal entries are +. Two sign patterns $\boldsymbol{A}_1, \boldsymbol{A}_2 \in Q_n$ are said to be permutation similar if $\boldsymbol{A}_2 = \boldsymbol{P}^{\mathrm{T}}\boldsymbol{A}_1\boldsymbol{P}$, for some permutation sign pattern $\boldsymbol{P}$.

A signature sign pattern is a square diagonal sign pattern matrix, each of whose diagonal entries is + or −. Two sign patterns $\boldsymbol{A}_1$, $\boldsymbol{A}_2 \in Q_n$ are said to be signature similar if $\boldsymbol{A}_2 = \boldsymbol{S}\boldsymbol{A}_1\boldsymbol{S}$, for some signature sign pattern $\boldsymbol{S}$.

A generalized permutation sign pattern is a sign pattern $\boldsymbol{A} \in Q_n$ that can be written as $\boldsymbol{A} = \boldsymbol{D}\boldsymbol{P}$, where $\boldsymbol{P}$ is a permutation sign pattern and $\boldsymbol{D}$ is a signature sign pattern.

Chapter 2

Sign Patterns That Allow Diagonalizability

This chapter content is our first research result on sign pattern to allow diagonalizability. In this chapter, firstly, we will consider the normal sign matrices that allow diagonalizability. Secondly, we will consider the necessary and sufficient conditions for an ordinary sign pattern to allow diagonalizability.

A sign pattern $\boldsymbol{A}$ is said to be normal if $\boldsymbol{A}\boldsymbol{A}^{\mathrm{T}}=\boldsymbol{A}^{\mathrm{T}}\boldsymbol{A}$. This chapter is organized as follows. In Section 2.1, the properties of normal sign patterns are discussed. In particular, normal sign matrices of order up to 3 are shown to allow diagonalizability. In Section 2.2, two necessary and sufficient conditions for a sign pattern that allow diagonalizability are obtained. Moreover, we also present two combinatorial necessary conditions for a sign pattern to allow diagonalizability.

2.1 Normal sign patterns

A square complex matrix $\boldsymbol{B}$ is said to be normal if $\boldsymbol{B}\boldsymbol{B}^{*}=\boldsymbol{B}^{*}\boldsymbol{B}$. It is well known that a square complex matrix is normal if and only if it is unitarily similar to a diagonal matrix [24]. For $\boldsymbol{A}\in Q_n$, if there exists a normal matrix $\boldsymbol{B}\in Q(\boldsymbol{A})$, we say that $\boldsymbol{A}$ allows normality.

Li et al. [31] investigated non-negative sign patterns that allow normality. The following three results are some basic properties of normal sign patterns.

Observation 2.1.1 [31] Let $\boldsymbol{A}$ be a non-negative sign pattern. If $\boldsymbol{A}$ allows normality, then $\boldsymbol{A}\boldsymbol{A}^{\mathrm{T}}=\boldsymbol{A}^{\mathrm{T}}\boldsymbol{A}$.

Proof Let $\boldsymbol{B}\in Q(\boldsymbol{A})$ be a normal matrix. Since $\boldsymbol{A}$ is non-negative, we have

$$\boldsymbol{A}\boldsymbol{A}^{\mathrm{T}} = \mathrm{sgn}(\boldsymbol{B}\boldsymbol{B}^{\mathrm{T}}) = \mathrm{sgn}(\boldsymbol{B}^{\mathrm{T}}\boldsymbol{B}) = \boldsymbol{A}^{\mathrm{T}}\boldsymbol{A}.$$

Proposition 2.1.2 Let $\boldsymbol{A}\in Q_n$ and suppose that $\boldsymbol{A}\boldsymbol{A}^{\mathrm{T}}=\boldsymbol{A}^{\mathrm{T}}\boldsymbol{A}$. Assume that rows i, j of $\boldsymbol{A}$ match, and the entries a_{ii}, a_{ij}, a_{ji} and a_{jj} are all non-zero. Then $a_{ij}=a_{ji}$.

Proof It suffices to consider the following two cases.

Case 1: $a_{ii}=a_{jj}$. If a_{ij} and a_{ji} are different, then rows i and j of $\boldsymbol{A}$ do not match, a contradiction. Hence, $a_{ij}=a_{ji}$.

Case 2: $a_{ii}=-a_{jj}$. Note that the hypotheses imply that the columns i, j of $\boldsymbol{A}$ match. If $a_{ij}\neq a_{ji}$, then the (i, j) entries of $\boldsymbol{A}\boldsymbol{A}^{\mathrm{T}}$ and $\boldsymbol{A}^{\mathrm{T}}\boldsymbol{A}$ are different, so that $\boldsymbol{A}\boldsymbol{A}^{\mathrm{T}}\neq\boldsymbol{A}^{\mathrm{T}}\boldsymbol{A}$, a contradiction. Hence, $a_{ij}=a_{ji}$.

Proposition 2.1.3 Let $A \in Q_n$ and suppose that $AA^T = A^TA$. Then rows i and j of A match, if and only if columns i and j of A match, for any i, $j \in \{1, 2, \cdots, n\}$.

The following three fundamental results are established in [10, 35].

Theorem 2.1.4 [10,37] Let $A \in Q_n$. If $c(A) = MR(A)$, then A allows diagonalizability.

Theorem 2.1.5 [10,37] Let $A \in Q_n$. If $c(A) \geqslant n-1$, then A allows diagonalizability.

Theorem 2.1.6 [37] Every combinatorially symmetric sign pattern A allows diagonalizability.

The following observation is useful in the sequel.

Lemma 2.1.7 The set of sign patterns that allow diagonalizability is closed under the following operations:

(1) negation;

(2) transposition;

(3) permutational similarity;

(4) signature similarity.

Proof Let $A \in Q_n$ be a sign pattern that allows diagonalizability and let $B \in Q(A)$ be a diagonalizable matrix. Clearly, $-B \in Q(-A)$ and $B^T \in Q(A^T)$ are diagonalizable. Hence, (i) and (ii) hold. Suppose that $P \in Q_n$ is a permutation sign pattern and $S \in Q_n$ is a signature sign pattern. Let P_1 be the (0, 1) matrix in $Q(P)$ and let S_1 be the (0, 1, −1) matrix in $Q(S)$. Note that $P_1^{-1} = P_1^T$ and $S_1^{-1} = S_1$. Since diagonalizability is invariant under similarity, $P_1^T B P_1$ is a diagonalizable matrix in $Q(P^TAP)$. Thus (iii) holds. Similarly, $S_1 B S_1$ is a diagonalizable matrix in $Q(SAS)$, which implies (iv).

Obviously, every sign pattern that allows normality also allows diagonalizability. As noted in [31], even for $A \in Q_2$, $AA^T = A^TA$ does not imply A allows normality. For instance, $A = \begin{bmatrix} + & + \\ 0 & + \end{bmatrix}$ satisfies $AA^T = A^TA$, but A does not allow normality. This sign pattern also provides an example of a sign pattern that allows diagonalizability, but does not allow normality.

By Theorem 2.1.6, every combinatorially symmetric sign pattern allows diagonalizability. We noticed that many sign patterns $\boldsymbol{A}$ satisfying $\boldsymbol{AA}^{\mathrm{T}} = \boldsymbol{A}^{\mathrm{T}}\boldsymbol{A}$, actually allow diagonalizability. This is not totally surprising since, the equation $\boldsymbol{AA}^{\mathrm{T}} = \boldsymbol{A}^{\mathrm{T}}\boldsymbol{A}$ implies that $\boldsymbol{A}$ enjoys a kind of symmetry in the following sense: row i of $\boldsymbol{A}$ is non-zero iff column i of $\boldsymbol{A}$ is non-zero for each i and the dot product of rows i and j of $\boldsymbol{A}$ could have the same sign as the dot product of columns i and j of $\boldsymbol{A}$, for all i and j. We are interested in determining to what extent the sign equation $\boldsymbol{AA}^{\mathrm{T}} = \boldsymbol{A}^{\mathrm{T}}\boldsymbol{A}$ can ensure that $\boldsymbol{A}$ allows diagonalizability.

For a 2×2 sign pattern $\boldsymbol{A}$, we can easily obtain the following result.

Theorem 2.1.8 Let $\boldsymbol{A} \in Q_2$. Suppose that $\boldsymbol{AA}^{\mathrm{T}} = \boldsymbol{A}^{\mathrm{T}}\boldsymbol{A}$, then $\boldsymbol{A}$ allows diagonalizability.

Proof If $c(\boldsymbol{A}) \geqslant 1$, then the conclusion follows from Theorem 2.1.5.

We now assume that $c(\boldsymbol{A}) = 0$, then $\boldsymbol{A}$ has no 1-cycle or 2-cycle. Hence, both of the diagonal entries of $\boldsymbol{A}$ are 0 and $\boldsymbol{A}$ is either strictly upper triangular or strictly lower triangular. It is readily seen that a non-zero strictly upper (or lower) triangular sign pattern $\boldsymbol{A} \in Q_2$ does not satisfy $\boldsymbol{AA}^{\mathrm{T}} = \boldsymbol{A}^{\mathrm{T}}\boldsymbol{A}$. The only strictly upper (or lower) triangular sign pattern $\boldsymbol{A} \in Q_2$ with $\boldsymbol{AA}^{\mathrm{T}} = \boldsymbol{A}^{\mathrm{T}}\boldsymbol{A}$ is $\boldsymbol{A} = 0$, which allows diagonalizability.

Now we are able to extend the preceding result to order 3.

Theorem 2.1.9 Let $\boldsymbol{A} \in Q_3$. If $\boldsymbol{AA}^{\mathrm{T}} = \boldsymbol{A}^{\mathrm{T}}\boldsymbol{A}$, then $\boldsymbol{A}$ allows diagonalizability.

Proof Suppose that $\boldsymbol{A} \in Q_3$ satisfies $\boldsymbol{AA}^{\mathrm{T}} = \boldsymbol{A}^{\mathrm{T}}\boldsymbol{A}$. If $c(\boldsymbol{A}) \geqslant 2$, then $\boldsymbol{A}$ allows diagonalizability by Theorem 2.1.5. We now assume that $c(\boldsymbol{A}) \leqslant 1$, then $\boldsymbol{A}$ has at most one non-zero diagonal entry, and $\boldsymbol{A}$ has no 2 - or 3 - cycles. Note that normal sign patterns as well as sign patterns that allow diagonalizability are closed under the operations mentioned in Lemma 2.1.7. It is known that a square matrix that has no cycle of length 2 or bigger is permutation similar to an upper triangular matrix[3]. By replacing $\boldsymbol{A}$ with $-\boldsymbol{A}$ if necessary, we may assume that the diagonal entries of $\boldsymbol{A}$ are non-negative. Further, by performing a permutation similarity on $\boldsymbol{A}$ if necessary, we may assume that $\boldsymbol{A}$ is in one of the following forms:

$$\boldsymbol{A}_1 = \begin{bmatrix} 0 & x & y \\ 0 & 0 & z \\ 0 & 0 & 0 \end{bmatrix},\ \boldsymbol{A}_2 = \begin{bmatrix} + & x & y \\ 0 & 0 & z \\ 0 & 0 & 0 \end{bmatrix},\ \boldsymbol{A}_3 = \begin{bmatrix} 0 & x & y \\ 0 & + & z \\ 0 & 0 & 0 \end{bmatrix},\ \boldsymbol{A}_4 = \begin{bmatrix} 0 & x & y \\ 0 & 0 & z \\ 0 & 0 & + \end{bmatrix},$$

where $x, y, z \in \{+, -, 0\}$.

For each $i = 1, 2, 3, 4$, if $\boldsymbol{A} = \boldsymbol{A}_i$, then it is readily verified that the condition $\boldsymbol{AA}^{\mathrm{T}} = \boldsymbol{A}^{\mathrm{T}}\boldsymbol{A}$ implies that $x = y = z = 0$. For example, in the case of $\boldsymbol{A} = \boldsymbol{A}_1$, by comparing the (1, 1) entries of $\boldsymbol{AA}^{\mathrm{T}}$ and $\boldsymbol{A}^{\mathrm{T}}\boldsymbol{A}$, we get $x^2 + y^2 = 0$, which implies $x = y = 0$. Also, for $\boldsymbol{A} = \boldsymbol{A}_1$, comparision of the (3, 3) entries of $\boldsymbol{AA}^{\mathrm{T}}$ and $\boldsymbol{A}^{\mathrm{T}}\boldsymbol{A}$ yields $y = z = 0$. Similarly, for $\boldsymbol{A} = \boldsymbol{A}_2$, comparison of the (3, 3) entries of $\boldsymbol{AA}^{\mathrm{T}}$ and $\boldsymbol{A}^{\mathrm{T}}\boldsymbol{A}$ yields $y = z = 0$, and hence comparison of the (2, 2) entries of $\boldsymbol{AA}^{\mathrm{T}}$ and $\boldsymbol{A}^{\mathrm{T}}\boldsymbol{A}$ yields $x = z = 0$.

Now that $x = y = z = 0$, $\boldsymbol{A}$ is in fact a diagonal sign pattern, and hence, $\boldsymbol{A}$ allows diagonalizability.

However, the following example shows that Theorem 2.1.9 cannot be extended to order 4.

Example 2.1.10 Let $\boldsymbol{A} = \begin{bmatrix} + & + & + & + \\ 0 & 0 & + & + \\ 0 & 0 & 0 & + \\ 0 & 0 & 0 & + \end{bmatrix}$. Then $\boldsymbol{AA}^{\mathrm{T}} = \boldsymbol{A}^{\mathrm{T}}\boldsymbol{A} = J$, where all the entries of J are +. Note that $c(\boldsymbol{A}) = 2$ and as $(a_{11})(a_{44})$ is the longest composite cycle of $\boldsymbol{A}$, with length 2 and $mr(\boldsymbol{A}) = MR(\boldsymbol{A}) = 3$. Therefore, for every $\boldsymbol{B} \in Q(\boldsymbol{A})$, 0 is an eigenvalue of $\boldsymbol{B}$, 0 has algebraic multiplicity 2 and geometric multiplicity 1. It follows that $\boldsymbol{B}$ is not diagonalizable. Thus $\boldsymbol{A}$ does not allow diagonalizability.

The following example provides a class of special sign patterns $\boldsymbol{A}$ such that $\boldsymbol{AA}^{\mathrm{T}} = \boldsymbol{A}^{\mathrm{T}}\boldsymbol{A}$ and $\boldsymbol{A}$ allows normality.

Example 2.1.11 Let $\boldsymbol{A} \in Q_n$ be a generalized permutation sign pattern. Then $\boldsymbol{AA}^{\mathrm{T}} = \boldsymbol{A}^{\mathrm{T}}\boldsymbol{A} = I_n$ and $\boldsymbol{A}$ allows real orthogonality. Indeed, the (0, 1, −1) matrix $\boldsymbol{B} \in Q(\boldsymbol{A})$ is a real orthogonal matrix, which is normal and diagonalizable. Hence, $\boldsymbol{A}$ allows diagonalizability.

We now give a characterization of generalized permutations.

Theorem 2.1.12 A sign pattern $\boldsymbol{A} \in Q_n$ satisfies $\boldsymbol{AA}^{\mathrm{T}} = \boldsymbol{I}_n$ iff $\boldsymbol{A}$ is a generalized permutation.

Proof Sufficiency is clear. We now prove necessity. Suppose that $\boldsymbol{A} \in Q_n$ satisfies $\boldsymbol{AA}^{\mathrm{T}} = \boldsymbol{I}_n$. Obviously, every matrix $\boldsymbol{B} \in Q(A)$ satisfies $\boldsymbol{BB}^{\mathrm{T}} \in Q(\boldsymbol{AA}^{\mathrm{T}}) = Q(\boldsymbol{I}_n)$ and hence $\boldsymbol{B}$ is non-singular. Thus each row of $\boldsymbol{A}$ is non-zero. Assume that the i th row of A has exactly $k \geqslant 2$ non-zero entries: a_{ij_1}, a_{ij_2}, $\cdots$, a_{ij_k}. Note that $AA^{\mathrm{T}} = I_n$ implies that each of the remaining $n-1$ rows of A would have formal dot product with the i - th row of A equal to 0. It follows that each of the remaining rows of A must have j_1, j_2, $\cdots$, j_k components equal to 0. Thus for every $B \in Q(A)$, the span of the $n-1$ rows of B other than the i th row has dimension at most $n-k$, so that rank $(B) \leqslant 1 + n - k < n$, contradicting the fact that B is non-singular. Therefore, each row of A contains precisely one non-zero entry. Since A requires non-singularity, these n non-zero entries of A must also be in distinct columns. It is then clear that A is a generalized permutation sign pattern.

From [37], we know that combinatorially symmetric sign patterns allow diagonalizability. From the preceding results, we know that some normal sign patterns also allow diagonalizability. However, it is easy to see that in general combinatorially symmetric sign patterns and normal sign patterns can be quite different. For instance, the matrix A in Example 2.1.10 is normal but not combinatorially symmetric, while the sign pattern $\begin{bmatrix} + & - \\ + & - \end{bmatrix}$ is combinatorially symmetric but not normal.

2.2 Necessary and sufficient conditions for allowing diagonalizability

By using the Jordan canonical form, it is easy to see that for any positive integer m, a square non-singular real matrix B is diagonalizable iff B^m is diagonalizable.

However, the following sign pattern $A = \begin{bmatrix} + & + & + & + & + \\ + & 0 & + & 0 & 0 \\ + & 0 & 0 & 0 & 0 \\ + & 0 & 0 & 0 & + \\ + & 0 & 0 & 0 & 0 \end{bmatrix}$ satisfies $A^2 = J$, so that A^2 allows diagonalizability, but A does not allow diagonalizability. By inspecting the digraph of A, we see that $\gamma_1 = a_{12}a_{23}a_{31}$ (or $\gamma_2 = a_{14}a_{45}a_{51}$) is a longest composite cycle of A. Thus $c(A) = 3$. Note that $mr(A) = MR(A) = 4$. It can be seen that every $B \in Q(A)$ has 0 as an eigenvalue with algebraic multiplicity

2 and geometric multiplicity 1. Thus A does not allow diagonalizability.

2.2.1 Non-singular matrix diagonally equivalent to a matrix with all distinct eigenvalues

2.2.1.1 2×2 matrices diagonally equivalent to matrices with two distinct eigenvalues

Two $n \times n$ matrices A and B over C are said to be diagonally equivalent if there are invertible diagonal matrices D_1 and D_2, such that $B = D_1AD_2$.

Matrices all of whose eigenvalues are distinct have many desirable properties, such as diagonalizability. Considerable research has been done on matrices with all distinct eigenvalues, see for example [23], [28], [30], [33], and [37]. In this section, we aim to identify matrices that are diagonally equivalent to matrices with no multiple eigenvalues.

Note that if D_1 and D_2 are invertible diagonal matrices of order n and A is any matrix of order n, then D_1AD_2 is similar to D_2D_1A. Thus in order to investigate the eigenvalues of D_1AD_2, it suffices to consider matrices of the form D_1A for invertible diagonal matrices D_1.

We denote the resultant (see [25], [34] or [40]) of two polynomials $f(x)$ and $g(x)$ by Res $(f(x), g(x))$. It is well known that $f(x)$ and $g(x)$ have no common zero (in an extension field that contains all the zeros of $f(x)$, $g(x)$) iff Res $(f(x), g(x)) \neq 0$. The discriminant (see [34] or [40]) of a polynomial $f(x)$, denoted discr $(f(x))$, is defined as the product of the squares of the pairwise differences of the roots of $f(x)$. It is also well known that a polynomial $f(x)$ has no multiple root iff Res $(f(x), f'(x)) \neq 0$, iff discr $(f(x)) \neq 0$. In fact, the discriminant of a monic polynomial $f(x)$ of degree n is given by

$$\mathrm{discr}(f(x)) = (-1)^{\frac{n(n-1)}{2}} \mathrm{Re}\, s(f(x), f'(x)).$$

Horn and Lopatin[23] gave an alternative method for finding the discriminant of the characteristic polynomial of a matrix A by computing the determinant of the moment matrix, whose (i, j) entry is the trace of A^{i+j-2}.

Theorem 2.2.1.1 A 2×2 matrix A is diagonally equivalent to a matrix with two distinct eigenvalues if and only if A is not strictly upper triangular or strictly lower triangular.

Proof We prove the equivalent statement that a 2×2 matrix $\boldsymbol{A}=\begin{bmatrix} a & b \\ c & d \end{bmatrix}$ is not diagonally equivalent to any matrix with two distinct eigenvalues if and only if $\boldsymbol{A}$ is strictly upper triangular or strictly lower triangular.

Clearly, if $\boldsymbol{A}$ is strictly upper (or lower) triangular, then every matrix diagonally equivalent to $\boldsymbol{A}$ is also strictly upper (or lower) triangular and thus would have 0 as the eigenvalue of multiplicity 2. Thus $\boldsymbol{A}$ is not diagonally equivalent to any matrix with two distinct eigenvalues.

We now prove the converse. Suppose that $\boldsymbol{A}=\begin{bmatrix} a & b \\ c & d \end{bmatrix}$ is not diagonally equivalent to any matrix with two distinct eigenvalues. Since a scalar multiple of a matrix preserves the property of the eigenvalues being distinct or otherwise, without loss of generality, we may restrict our attention to matrices of the form $\boldsymbol{D}_2\boldsymbol{A}$, where $\boldsymbol{D}_2$ is a diagonal matrix of the form $\boldsymbol{D}_2=\begin{bmatrix} 1 & 0 \\ 0 & x \end{bmatrix}$, $x\neq0$. The characteristic polynomial (in t) of $\boldsymbol{D}_2\boldsymbol{A}$ is

$$t^2-(a+dx)\,t+x\boldsymbol{D},$$

where $\boldsymbol{D}=ad-bc$ denotes the determinant of $\boldsymbol{A}$. Since $\boldsymbol{D}_2\boldsymbol{A}$ has a multiple eigenvalue for every non-zero x, the discriminant of the above characteristic polynomial is equal to 0, namely,

$$(a+dx)^2-4x\boldsymbol{D}=0, \text{or } d^2x^2+(2ad-4\boldsymbol{D})x+a^2=0$$

for all non-zero values of x. For the polynomial $d^2x^2+(2ad-4\boldsymbol{D})\,x+a^2$ to have infinitely many roots, all the coefficients must be zero. Thus, $d^2=0$, $a^2=0$ and $2ad-4\boldsymbol{D}=0$. It follows that $a=0$, $d=0$ and $\boldsymbol{D}=0$. Consequently, $bc=ad-\boldsymbol{D}=0$. We now have $a=d=0$ and $(c=0$ or $b=0)$, namely, $\boldsymbol{A}$ is strictly upper triangular or $\boldsymbol{A}$ is strictly lower triangular.

The above theorem can be rephrased in terms of the principal minors as follows.

Corollary 2.2.1.2 A 2×2 matrix $\boldsymbol{A}$ is not diagonally equivalent to any matrix with two distinct eigenvalues if and only if $\det \boldsymbol{A}\neq0$ and each diagonal entry of $\boldsymbol{A}$ is zero, if and only if each principal minor of $\boldsymbol{A}$ is zero.

The following result follows immediately from the above theorem.

Corollary 2. 2. 1. 3 Every 2×2 non-singular matrix is diagonally equivalent a matrix with two distinct eigenvalues.

2. 2. 1. 2 3×3 matrices diagonally equivalent to matrices with 3 distinct eigenvalues

The main result of this section is the following.

Theorem 2. 2. 1. 4 A 3×3 complex matrix $\boldsymbol{A}$ is not diagonally equivalent to any matrix with 3 distinct eigenvalues if and only if $\det \boldsymbol{A}\neq 0$ and every principal minor of $\boldsymbol{A}$ of order 2 is zero.

In order to prove Theorem 2. 2. 1. 4, we need the following lemma, which can be proved by induction on the number of variables following the proof of Theorem 2. 19 in section 2. 12 of [25]. For convenience, we define the degree of the zero polynomial to be $-\infty$, which is naturally considered to be less than any integer.

Lemma 2. 2. 1. 5 Let $f(x_1, x_2, \cdots, x_n)$ be a polynomial over a field $\boldsymbol{F}$ with degree at most m in each variable x_i. Let $S_i(1\leqslant i\leqslant n)$ be a subset of $\boldsymbol{F}$ with at least $m+1$ elements. Suppose that $f(c_1, c_2, \cdots, c_n)=0$ for all $(c_1, c_2, \cdots, c_n)\in S_1\times S_2\times\cdots\times S_n$. Then $f(x_1, x_2, \cdots, x_n)$ is the zero polynomial, namely, all of the coefficients of $f(x_1, x_2, \cdots, x_n)$ are zeros.

Proof of Theorem 2. 2. 1. 4 Consider a 3×3 complex matrix

$$\boldsymbol{A}=\begin{bmatrix} a & b & c \\ d & e & f \\ g & h & k \end{bmatrix}.$$

For an invertible diagonal matrix $\boldsymbol{D}_3=diag(x, y, z)$, the characteristic polynomial of $\boldsymbol{D}_3\boldsymbol{A}$ is

$$p(t)=t^3-(ax+ey+kz)\ t^2+(xyM_1+yzM_2+xzM_3)t-xyzD,$$

where $D=\det \boldsymbol{A}=aek+bfg+cdh-afh-bdk-ceg$, while $M_1=(ae-bd)$, $M_2=(ek-fh)$ and $M_3=(ak-cg)$ are the principal minors of $\boldsymbol{A}$ of order 2. Hence,

$$p'(t)=3t^2-2(ax+ey+kz)\ t+(xyM_1+yzM_2+xzM_3),$$

With the help of Maple, we find that

$Res(p(t),p'(t))=-a^2M_1{}^2x^4y^2-(2a^2M_3M_1+4a^3D)x^4yz-a^2M_3{}^2x^4z^2+(4M_1{}^3-2aeM_1{}^2)x^3y^3+(12M_1{}^2M_3-2akM_1{}^2-18aDM_1+12a^2eD-2a^2M_1M_2-4aeM_1M_3)x^3y^2z+(12M_1M_3{}^2-4akM_3M_1-2a^2M_3M_2-18aDM_3-2aeM_3{}^2+$

$12a^2kD)x^3yz^2 + (4M_3{}^3 - 2akM_3{}^2)x^3z^3 - e^2M_1{}^2x^2y^4 + (-18eDM_1 - 4aeM_1M_2 - 2e^2M_1M_3 - 2ekM_1{}^2 + 12M_1{}^2M_2 + 12ae^2D)x^2y^3z + (24M_1M_2M_3 - 4aeM_3M_2 - 4akM_1M_2 - k^2M_1{}^2 - 4ekM_1M_3 + 27D^2 - e^2M_3{}^2 - 18eDM_3 - 18kDM_1 - 18aDM_2 + 24aekD - a^2M_2{}^2)x^2y^2z^2 + (-18kDM_3 - 2k^2M_3M_1 - 4akM_3M_2 + 12M_2M_3{}^2 - 2ekM_3{}^2 + 12ak^2D)x^2yz^3 - k^2M_3{}^2x^2z^4 + (4e^3D - 2e^2M_1M_2)xy^4z + (12e^2kD - 2e^2M_2M_3 - 18eDM_2 + 12M_1M_2{}^2 - 2aeM_2{}^2 - 4eM_1kM_2) \times xy^3z^2 + (-18kDM_2 + 12k^2eD + 12M_2{}^2M_3 - 4ekM_3M_2 - 2akM_2{}^2 - 2k^2M_2M_1)xy^2z^3 + (-2k^2M_3M_2 + 4k^3D)xyz^4 - e^2M_2{}^2y^4z^2 + (4M_2{}^3 - 2ekM_2{}^2)y^3z^3 - k^2M_2{}^2y^2z^4.$

The sufficiency of the theorem is clear. Indeed, suppose that det $\boldsymbol{A} = 0$ and every principal minor of $\boldsymbol{A}$ of order 2 is zero. Then $D = M_1 = M_2 = M_3 = 0$. Hence, for every non-singular diagonal matrix $\boldsymbol{D}_3$, the characteristic polynomial of $\boldsymbol{D}_3\boldsymbol{A}$ is $p(t) = t^3 - (ax + ey + kz)t^2$, which has 0 as a multiple root. Thus $\boldsymbol{A}$ is not diagonally equivalent to any matrix with 3 distinct eigenvalues.

We now prove the necessity. Assume that the 3×3 matrix $\boldsymbol{A}$ is not diagonally equivalent to any matrix with 3 distinct eigenvalues. Then $\boldsymbol{D}_3\boldsymbol{A} = diag(x, y, z)A$ has a multiple eigenvalue for all positive integers x, y and z in the set $S = \{1, 2, 3, 4, 5\}$. Then Res $(p(t), p'(t)) = 0$ for all $(x, y, z) \in S^3$. By Lemma 1.3.1.5, all the coefficients of the monomials in x, y and z in Res $(p(t), p'(t))$ are equal to zero.

Assume that $a \neq 0$. By inspecting the coefficient of x^4y^2 in Res $(p(t), p'(t))$, we get $-a^2M_1^2 = 0$. Hence, $M_1 = ae - bd = 0$. Similarly, by inspecting the coefficient of x^4z^2, we get $M_3 = 0$. Then by considering the coefficient of x^4yz, we have

$$4a^3D - 2a^2M_1M_3 = 0.$$

It then follows from $a \neq 0$ and $M_1 = 0$ that $D = 0$. Now, all the terms of the coefficient of $x^2y^2z^2$ except possibly $-2a^2M_2^2$ are obviously zero since they involve M_1, M_2 or D as a factor. It follows that $-2a^2M_2^2 = 0$ and hence, $M_2 = 0$. Thus, $D = \det\boldsymbol{A} = 0$ and every principal minor of $\boldsymbol{A}$ of order 2 is zero.

Similarly, if $e \neq 0$, then by inspecting the coefficients of y^4z^3, x^2y^4, xy^4z and $x^2y^2z^2$, we get $M_2 = 0$, $M_1 = 0$, $D = 0$ and $M_3 = 0$.

Also, if $k \neq 0$, then by inspecting the coefficients of y^2z^4, x^2z^4, xyz^4 and

$x^2y^2z^2$, we get $M_3=0$, $M_2=0$, $D=0$ and $M_1=0$.

We now consider the case of $a=e=k=0$. By inspecting the coefficient of x^3y^3, we get $4M_1{}^3=2eM_1^2a=0$ and hence, $M_1=0$. Similarly, by inspecting the coefficients of y^3z^3 and x^3z^3, we get $M_2=0$ and $M_3=0$. It follows that the first term in the coefficient of $x^2y^2z^2$ is 0 since it involves a factor of M_1 while all the other terms except possibly $27D^2$ are 0 since they involve a, e or k as a factor. Thus we have $27D^2=0$ and hence, $D=0$.

Therefore, for every 3×3 matrix $\boldsymbol{A}$ that is not diagonally equivalent to any matrix with three distinct eigenvalues, we have $\det\boldsymbol{A}=0$ and every principal minor of $\boldsymbol{A}$ of order 2 is 0.

Observe that as indicated in the proof of Theorem 2.2.1.4, if a 3×3 complex matrix $\boldsymbol{A}$ is diagonally equivalent to a matrix with 3 distinct eigenvalues, then there is diagonal matrix $\boldsymbol{D}_3$ with each diagonal entry a suitable integer between 1 and 5 such that $\boldsymbol{D}_3\boldsymbol{A}$ has 3 distinct eigenvalues.

More generally, it can be seen that for an $n\times n$ matrix $\boldsymbol{A}$, the discriminant of the characteristic polynomial of $diag(x_1, x_2, \cdots, x_n)$, $\boldsymbol{A}$ has degree less than $2n$ in each variable x_i. If an $n\times n$ complex matrix $\boldsymbol{A}$ is diagonally equivalent to a matrix with n distinct eigenvalues, then there is diagonal matrix $\boldsymbol{D}_n$ whose diagonal entries are suitable integers between 1 and $2n$ such that $\boldsymbol{D}_n\boldsymbol{A}$ has n distinct eigenvalues.

Theorem 2.2.1.4 may be restated as follows.

Theorem 2.2.1.6 A 3×3 complex matrix $\boldsymbol{A}$ is diagonally equivalent to a matrix with 3 distinct eigenvalues if and only if $\det\boldsymbol{A}\neq0$ or a principal minor of $\boldsymbol{A}$ of order 2 is non-zero.

As an immediate consequence of Theorem 2.2.1.6, we have the following result.

Corollary 2.2.1.7 Every nonsingular 3×3 complex matrix $\boldsymbol{A}$ is diagonally equivalent to a matrix with 3 distinct eigenvalues.

Example 2.2.1.8 The following 3×3 matrices are not diagonally equivalent to any matrix with 3 distinct eigenvalues:

$$\begin{bmatrix}1&1&-1\\2&2&-2\\3&3&-3\end{bmatrix},\ \begin{bmatrix}1&1&0\\1&1&1\\0&0&0\end{bmatrix},\ \begin{bmatrix}0&0&1\\1&1&1\\0&0&0\end{bmatrix},\ \begin{bmatrix}0&1&2\\0&0&3\\0&0&0\end{bmatrix}.$$

It can be easily verified that a 3×3 matrix $\boldsymbol{A}$ with all entries non-zero is not diagonally equivalent to any matrix with 3 distinct eigenvalues if and only if $rank\ \boldsymbol{A} = 1$. In this case, $\boldsymbol{A}$ is diagonally equivalent to a matrix all of whose entries in the first row or column are equal to 1.

2.2.1.2 4 × 4 matrices and beyond

We conjecture that Theorem 2.2.1.4 holds for all orders $n \geqslant 2$.

Conjecture 2.2.1.9 For all $n \geqslant 2$, an $n \times n$ complex matrix $\boldsymbol{A}$ is diagonally equivalent to a matrix with n distinct eigenvalues if and only if $\det \boldsymbol{A} \neq 0$ or a principal minor of $\boldsymbol{A}$ of order $n-1$ is non-zero.

Evidently, if an $n \times n$ matrix $\boldsymbol{A}$ satisfies that every principal minor of $\boldsymbol{A}$ of order $\geqslant n-1$ is 0, then every matrix diagonally equivalent to $\boldsymbol{A}$ also has this property and hence would have 0 as a multiple eigenvalue.

Conjecture 2.2.1.9 claims that if an $n \times n$ complex matrix $\boldsymbol{A}$ satisfies the condition that for every invertible diagonal matrix $\boldsymbol{D}_n$, $\boldsymbol{D}_n\boldsymbol{A}$ has a multiple eigenvalue, then 0 is a multiple eigenvalue of $\boldsymbol{D}_n\boldsymbol{A}$ for every invertible diagonal matrix $\boldsymbol{D}_n$.

A weaker version of the above conjecture is the following.

Conjecture 2.2.1.10 For all positive integers n, every non-singular $n \times n$ complex matrix $\boldsymbol{A}$ is diagonally equivalent to a matrix with n distinct eigenvalues.

To demonstrate the difficulties that we encounter when we try to resolve the above conjecture for higher orders, let us have a glimpse of what happens when $n = 4$.

Consider a 4×4 matrix

$$\boldsymbol{B} = \begin{bmatrix} a & b & c & d \\ e & f & g & h \\ i & j & k & l \\ m & n & p & q \end{bmatrix}$$

and a generic invertible diagonal matrix $\boldsymbol{D}_4 = diag\ (w, x, y, z)$.

The characteristic polynomial of $\boldsymbol{D}_4\boldsymbol{B}$ is

$$p_4(t) = t^4 - (aw + fx + ky + qz)\, t^3 + (wxM_1 + wyM_2 + wzM_3 + xyM_4 + xzM_5 + yzM_6)\, t^2 - (wxyK_1 + xyzK_2 + yzwK_3 + zwxK_4)\, t - wxyzD$$

where M_1, M_2, $\cdots$, M_6 are the 2×2 principal minors of $\boldsymbol{B}$, K_1, K_2, K_3, K_4 are the 3×3 principal minors of $\boldsymbol{B}$, and $D=\det\boldsymbol{B}$.

We show a small fraction of the terms indiscr $(p_4(t))$ obtained with the help of Maple, by denoting $p_4(t)$ as poly 4 and using the command sort (collect (discrim (poly 4, t), [w, x, y, z], distributed), [w, x, y, z]);

discr $(p_4(t)) = a^2M_1{}^2K_1{}^2w^6x^4y^2 + (-4a^2DM_1{}^3 + 2a^2K_4M_1{}^2K_1)w^6x^4yz + a^2K_4{}^2M_1{}^2w^6x^4z^2 + (-4a^3K_1{}^3 + 2a^2M_1M_2K_1{}^2)w^6x^3y^3 + (18a^3DM_1K_1 - 12a^3K_4K_1{}^2 - 12a^2DM_1{}^2M_2 ++ 4a^2K_4M_1M_2K_1 + 2a^2M_1{}^2K_1K_3 + 2a^2M_1M_3K_1{}^2)w^6x^3y^2z + (18a^3DK_4M_1 - 12a^3K_4{}^2K_1 - 12a^2DM_1{}^2M_3 + 2a^2K_4{}^2M_1M_2 + 2a^2K_4M_1{}^2K_3 + 4a^2K_4M_1M_3K_1)w^6x^3yz^2 + (-4a^3K_4{}^3 + 2a^2K_4{}^2M_1M_3)w^6x^3z^3 + a^2M_2{}^2K_1{}^2w^6x^2y^4 + (18a^3DM_2K_1 - 12a^3K_1{}^2K_3 - 12a^2DM_1M_2{}^2 + 2a^2K_4M_2{}^2K_1 + 4a^2M_1M_2K_1K_3 + 2a^2M_2M_3K_1{}^2)w^6x^2y^3z + (-27a^4D^2 + 18a^3DK_4M_2 + 18a^3DM_1K_3 + 18a^3DM_3K_1 - 24a^3K_4K_1K_3 - 24a^2DM_1M_2M_3 + a^2K_4{}^2M_2{}^2 + 4a^2K_4M_1M_2K_3 + 4a^2K_4M_2M_3K_1 + a^2M_1{}^2K_3{}^2 + 4a^2M_1M_3K_1K_3 + a^2M_3{}^2K_1{}^2)w^6x^2y^2z^2 + (18a^3DK_4M_3 - 12a^3K_4{}^2K_3 - 12a^2DM_1M_3{}^2 + 2a^2K_4{}^2M_2K_3 + 4a^2K_4M_1M_3K_3 + 2a^2K_4M_3{}^2K_1)w^6x^2yz^3 + a^2K_4{}^2M_3{}^2M_2K_3w^6x^2z^4 + (-4a^2DM_2{}^3 + 2a^2M_2{}^2K_1K_3)w^6xy^4z + (18a^3DM_2K_3 - 12a^3K_1K_3{}^2 - 12a^2DM_2{}^2M_3 + 2a^2K_4M_2{}^2K_3 + 2a^2M_1M_2K_3{}^2 + 4a^2M_2M_3K_1K_3)w^6xy^3z^2 + (18a^3DM_3K_3 - 12a^3K_4K_3{}^2 - 12a^2DM_2M_3{}^2 + 4a^2K_4M_2M_3K_3 + 2a^2M_1M_3K_3{}^2 + 2a^2M_3{}^2K_1K_3)w^6xy^2z^3 + (-4a^2DM_3{}^3 + 2a^2K_4M_3{}^2K_3)w^6xyz^4 + a^2M_2{}^2K_3{}^2w^6y^4z^2 + (-4a^3K_3{}^3 + 2a^2M_2M_3K_3{}^2)w^6y^3z^3 + a^2M_3{}^2K_3{}^2w^6y^2z^4 + (2afM_1{}^2K_1{}^2 - 4M_1{}^3K_1{}^2)w^5x^5y^2 + (-8afDM_1{}^3 + 4afK_4M_1{}^2K_1 + 16DM_1{}^4 - 8K_4M_1{}^3K_1)w^5x^5yz + (2afK_4{}^2M_1{}^2 - 4K_4{}^2M_1{}^3)w^5x^5z^2 + \cdots + (2fqK_2{}^2M_5{}^2 - 4K_2{}^2M_5{}^3)x^5y^2z^5 + k^2K_2{}^2M_4{}^2x^4y^6z^2 + (-12fk^2K_2{}^3 + 4fkK_2{}^2M_4M_6 + 2k^2K_2{}^2M_4M_5 + 2kqK_2{}^2M_4{}^2 + 18kK_2{}^3M_4 - 12K_2{}^2M_4{}^2M_6)x^4y^5z^3 + (f^2K_2{}^2M_6{}^2 - 24fkqK_2{}^3 + 4fkK_2{}^2M_5M_6 + 4fqK_2{}^2M_4M_6 + k^2K_2{}^2M_5{}^2 + 4kqK_2{}^2M_4M_5 + q^2K_2{}^2M_4{}^2 + 18fK_2{}^3M_6 + 18kK_2{}^3M_5 + 18qK_2{}^3M_4 - 24K_2{}^2M_4M_5M_6 - 27K_2{}^4)x^4y^4z^4 + (-12fq^2K_2{}^3 + 4fqK_2{}^2M_5M_6 + 2kqK_2{}^2M_5{}^2 + 2q^2K_2{}^2M_4M_5 + 18qK_2{}^3M_5 - 12K_2{}^2M_5{}^2M_6) \times x^4y^3z^5 + q^2K_2{}^2M_5{}^2x^4y^2z^6 + (-4k^3K_2{}^3 + 2k^2K_2{}^2M_4M_6)x^3y^6z^3 + (2fkK_2{}^2M_6{}^2 - 12k^2qK_2{}^3 -+ 2k^2K_2{}^2M_5M_6 + 4kqK_2{}^2M_4M_6 + 18kK_2{}^3M_6 - 12K_2{}^2M_4M_6^2)x^3y^5z^4 + (2fqK_2{}^2M_6{}^2 - 12kq^2K_2{}^3 + 4kqK_2{}^2M_5M_6 + 2q^2K_2{}^2M_4M_6 + 18qK_2{}^3M_6 - 12K_2{}^2M_5M_6^2)x^3y^4z^5 + (-4q^3K_2{}^3 + 2q^2K_2{}^2M_5M_6)x^3y^3z^6 + k^2K_2{}^2M_6{}^2x^2y^6z^4 + (2kqK_2{}^2M_6{}^2 - 4K_2{}^2M_6^3)x^2y^5z^5 +$

$q^2K_2{}^2M_6{}^2x^2y^4z^6$.

However, the entire expression for discr $(p_4(t))$, a polynomial in w, x, y and z of degree at most 6 in each variable, is about 34 pages long in Maple output and is by far more complicated than its counterpart for $n = 3$. The coefficient of $w^3x^3y^3z^3$ alone involves more than 200 terms, one of which is $256D^3$.

Assume that $\boldsymbol{B}$ is not diagonally equivalent to any matrix with 4 distinct eigenvalues. Then all the coefficients (of the monomials in w, x, y and z) in discr $(p_4\ (t))$ are equal to 0.

Let us consider the special case that all diagonal entries of $\boldsymbol{B}$ are non-zero. Replacing $\boldsymbol{B}$ by a suitable matrix diagonally equivalent to $\boldsymbol{B}$ if necessary, we may assume that all the diagonal entries of $\boldsymbol{B}$ are equal to 1. Since $a \neq 0$, by inspecting the coefficients of $w^6x^4y^2$ and $w^6x^3y^3$, and considering the two cases of $M_1 \neq 0$ or $M_1 = 0$, we can see that $K_1 = 0$. Similarly, by using $f \neq 0$, $k \neq 0$, $q \neq 0$ and inspecting the coefficients of certain monomials in discr ($(p_4\ (t))$) whose degree in one variable is 6, it can be easily seen that $K_2 = K_3 = K_4 = 0$. If $M_1 \neq 0$, then an inspection of the coefficient of w^6x^4yz reveals that $D = 0$. If $M_1 = 0$, then note that with the exception of $-27a^4D^2$, every term of the coefficient of $w^6x^2y^2z^2$ involves M_1 or one of K_1, K_2, K_3 and K_4 as a factor and hence, is equal to 0. Thus, $27a^4D^2 = 0$. It follows that $D = 0$.

Showing that $K_1 = K_2 = K_3 = K_4 = 0$ and $D = \det B = 0$ for all 4×4 matrices $\boldsymbol{B}$ that are not diagonally equivalent to any matrix with 4 distinct eigenvalues remains an interesting challenge.

In 1991, C. R. Johnson[26] proved that for any sign pattern $\boldsymbol{A}$ and any integer k with $mr(\boldsymbol{A}) \leqslant k \leqslant MR(\boldsymbol{A})$, there exists a real matrix $\boldsymbol{B} \in Q(\boldsymbol{A})$ such that $rank\ \boldsymbol{B} = k$. We call this result Johnson's intermediate rank theorem, a proof of which can also be found in [37].

In fact, Conjecture 2. 2. 1. 10 is right, and Choi presented the answer[5]. We will discuss it in the next part.

2. 2. 2 Sign patterns that allow diagonalizabilty

The following result can be seen by combining the results from next two papers.

Theorem 2. 2. 2. 1[5] Let $\boldsymbol{A}$ be an $n \times n$ non-singular matrix. Then there exists an invertible $n \times n$ diagonal matrix $\boldsymbol{D}$ such that $\boldsymbol{DA}$ has n distinct eigenvalues.

The following stronger result can be shown by slight modification of the proof of the preceding theorem in [5] or by combining Theorem 2.2.2.1 with the discriminant approach in [13].

Theorem 2.2.2.2[13,5] Let $\boldsymbol{A}$ be a $n \times n$ non-singular matrix. Then there exists an invertible $n \times n$ diagonal matrix $\boldsymbol{D}$ with positive diagonal entries such that $\boldsymbol{DA}$ has n distinct eigenvalues.

Recall that a square real matrix $\boldsymbol{B}$ is said to be rank principal if $\boldsymbol{B}$ has an invertible principal submatrix that has the same rank as $\boldsymbol{B}$.

The preceding theorem enables us to establish the following theorem that says a sign pattern allows diagonalizability if it allows rank principality.

Theorem 2.2.2.3 A sign pattern $\boldsymbol{A} \in Q_n$ allows diagonalizability if and only if there exists a real matrix $\boldsymbol{B} \in Q(\boldsymbol{A})$ with rank $B = k$, such that B has a non-singular $k \times k$ principal submatrix.

Proof Sufficiency. Suppose that $\boldsymbol{B} \in Q(\boldsymbol{A})$ satisfies rank $\boldsymbol{B} = k$ and $\boldsymbol{B}$ has an invertible $k \times k$ principal submatrix. Replacing $\boldsymbol{B}$ by $\boldsymbol{P}^{\mathrm{T}}\boldsymbol{BP}$ for some suitable permutation matrix $\boldsymbol{P}$ if necessary, we may assume that the $k \times k$ leading principal submatrix of $\boldsymbol{B}$ is invertible. Write

$$\boldsymbol{B} = \begin{bmatrix} \boldsymbol{B}_{11} & \boldsymbol{B}_{12} \\ \boldsymbol{B}_{21} & \boldsymbol{B}_{22} \end{bmatrix}.$$

where $\boldsymbol{B}_{11}$ is a $k \times k$ invertible submatrix of $\boldsymbol{B}$. By Theorem 2.2.2.2, there exists an invertible $k \times k$ diagonal matrix $\boldsymbol{D}_0$ with positive diagonal entries such that $\boldsymbol{D}_0\boldsymbol{B}_{11}$ has k distinct non-zero eigenvalues. Let

$\boldsymbol{D}_1 = diag(\boldsymbol{D}_0, \ x_1, \ \cdots, \ x_{n-k})$ and $\boldsymbol{D}_2 = diag(I_k, \ y_1, \ y_2, \ \cdots, \ y_{n-k})$.

where $x_1, x_2, \cdots, x_{n-k}, y_1, y_2, \cdots, y_{n-k}$ are positive numbers. Consider

$$\boldsymbol{D}_1\boldsymbol{BD}_2 = \begin{bmatrix} \boldsymbol{D}_0\boldsymbol{B}_{11} & \boldsymbol{B}'_{12} \\ \boldsymbol{B}'_{21} & \boldsymbol{B}'_{22} \end{bmatrix} \in Q(\boldsymbol{A}).$$

By choosing $x_1, x_2, \cdots, x_{n-k}, y_1, y_2, \cdots, y_{n-k}$ to be sufficiently small positive numbers, we can make all the entries of $\boldsymbol{B}'_{12}$, $\boldsymbol{B}'_{21}$ and $\boldsymbol{B}'_{22}$ to be arbitrarily small. Since the eigenvalues of a matrix continuously depend on the entries of the matrix, we assume that $\boldsymbol{D}_1$ and $\boldsymbol{D}_2$ are chosen so that $\boldsymbol{D}_1\boldsymbol{BD}_2$ has k distinct non-zero eigenvalues. Since the the leading $k \times k$ principal submatrix of $\boldsymbol{D}_1\boldsymbol{BD}_2$ is non-singular and rank $(D_1BD_2)=k)$, the non-singular matrix

$$S=\begin{bmatrix} I_k & 0 \\ -B'_{21}(D_0B_{11})^{-1} & I_{n-k}\end{bmatrix} \text{ satisfies } SD_1BD_2=\begin{bmatrix} D_0B_{11} & B'_{12} \\ 0 & 0\end{bmatrix}.$$

Furthermore,

$$SD_1BD_2S^{-1}=\begin{bmatrix} B'_{11} & B'_{12} \\ 0 & 0\end{bmatrix}.$$

Since D_1BD_2, and hence $SD_1BD_2S^{-1}$, has k distinct non-zero eigenvalues, we see that B'_{11} has k distinct non-zero eigenvalues. Obviously, the algebraic multiplicity and geometric multiplicity of the eigenvalue 0 of $SD_1BD_2S^{-1}$ are both equal to $n-k$. Therefore, $SD_1BD_2S^{-1}$ is diagonalizable. Thus $D_1BD_2\in Q(A)$ is diagonalizable, namely, A allows diagonalizability.

We now prove necessity. Suppose that there exists a diagonalizable matrix $B\in Q(A)$. Let $k=rank(B)$. Then B has k non-zero eigenvalues (counting algebraic multiplicities). The characteristic polynomial of B is

$$p_B(t)=t^n-E_1(B)t^{n-1}+E_2(B)t^{n-2}-\cdots+(-1)^nE_n(B),$$

where $E_i(B)$ is the sum of all i − by − i principal minors of B, $i=1, 2, \cdots, n$, see [24].

Suppose there is no non-zero $k\times k$ principal minor of B. Then $E_k(B)=0$. Since rank $B=k$, we also have $E_i(B)=0$ for all $i>k$. Therefore, the characteristic polynomial of B has the form

$$\begin{aligned} p_B(t) &= t^n-E_1(B)\,t^{n-1}+E_2(B)\,t^{n-2}-\cdots+(-1)^{k-1}E_{k-1}(B)\,t^{n-k+1} \\ &= t^{n-k+1}(t^{k-1}-E_1(B)\,t^{k-2}+E_2(B)\,t^{n-3}-\cdots+(-1)^{k-1}E_{k-1}(B)). \end{aligned}$$

It follows that B has at least $n-k+1$ zero eigenvalues, and hence, B has at most $k-1$ non-zero eigenvalues, contradicting the fact that B has k non-zero eigenvalues. Therefore, B has a non-zero $k\times k$ principal minor, or equivalently, B has an invertible $k\times k$ principal submatrix.

It clear that every diagonalizable matrix B satisfies $rank(B)=\text{rank}(B^2)$. Thus every sign pattern that allows diagonalizability also allows this rank stability property. We now show that the converse is also true. Thus we obtain another necessary and sufficient condition for a sign pattern to allow diagonalizability.

Theorem 2.2.2.4 Let $A\in Q_n$. Then A allows diagonalizability iff there is a *matrix* $B \in Q(\boldsymbol{A})$ such that $rank(\boldsymbol{B}) = rank(\boldsymbol{B}^2)$.

Proof Necessity is clear. We now prove sufficiency. Suppose that there is a matrix $\boldsymbol{B} \in Q(\boldsymbol{A})$ such that $rank(\boldsymbol{B}) = rank(\boldsymbol{B}^2)$. Let $rank(\boldsymbol{B}) = k$. The condition $rank(\boldsymbol{B}) = rank(\boldsymbol{B}^2)$ implies that every Jordan block (if any) of $\boldsymbol{B}$ corresponding to 0 in the Jordan canonical form of $\boldsymbol{B}$ is 1×1 and is of rank 0. It follows that the sum of the orders of the Jordan blocks of $\boldsymbol{B}$ corresponding to non-zero eigenvalues is $k = rank(\boldsymbol{B})$, and hence, $\boldsymbol{B}$ has precisely k non-zero eigenvalues (counting algebraic multiplicities) and $n-k$ eigenvalues equal to 0. Since $E_k(\boldsymbol{B})$ is equal to the symmetric sum of the products of k eigenvalues of $\boldsymbol{B}$ and this sum has a unique non-zero term, we have $E_k(\boldsymbol{B}) \neq 0$. Thus $\boldsymbol{B}$ has a non-singular $k \times k$ principal submatrix, namely, $\boldsymbol{B}$ is rank principal. By Theorem 2.2.2.3, $\boldsymbol{A}$ allows diagonalizability.

Suppose that a sign pattern $\boldsymbol{A}$ allows diagonalizability and let $\boldsymbol{B} \in Q(\boldsymbol{A})$ be a diagonalizable matrix. Then

$$mr(\boldsymbol{A}) \leqslant rank(\boldsymbol{B}) = rank(\boldsymbol{B}^2) \leqslant MR(\boldsymbol{A}^2).$$

Thus we have the following combinatorial necessary condition for a sign pattern to allow diagonalizability.

Corollary 2.2.2.5 If a sign pattern $\boldsymbol{A}$ allows diagonalizability, then

$$mr(\boldsymbol{A}) \leqslant MR(\boldsymbol{A}^2).$$

Example 2.2.2.6 Let

$$\boldsymbol{A} = \begin{bmatrix} + & + & + & + \\ - & + & + & + \\ 0 & 0 & 0 & + \\ 0 & 0 & 0 & 0 \end{bmatrix}. \text{ Then } \boldsymbol{A}^2 = \begin{bmatrix} \# & + & + & + \\ - & \# & \# & \# \\ 0 & 0 & 0 & 0 \\ 0 & 0 & 0 & 0 \end{bmatrix}.$$

Since $mr(\boldsymbol{A}) = 3 > MR(\boldsymbol{A}^2) = 2$, by Corollary 2.2.2.5, $\boldsymbol{A}$ does not allow diagonalizability.

Using different notation, Eschenbach and Johnson [9] gave the necessary condition $n - c(\boldsymbol{A}) \leqslant n - mr(\boldsymbol{A})$ for a sign pattern matrix $\boldsymbol{A} \in Q_n$ to allow diagonalizability, which can be restated in a more convenient form as follows. We provide a new proof.

Theorem 2.2.2.7[10] Suppose that a sign pattern $\boldsymbol{A}$ allows diagonalizability. Then

$mr(\boldsymbol{A}) \leqslant c(\boldsymbol{A})$.

Proof By Theorem 2. 2. 2. 3, there is a rank principal matrix $\boldsymbol{B} \in Q(\boldsymbol{A})$. Let $k = rank(\boldsymbol{B})$. Since $\boldsymbol{B}$ is rank principal, it has a non-singular $k \times k$ principal submatrix $\boldsymbol{C}$. The fact that det $\boldsymbol{C} \neq 0$ implies that $\boldsymbol{C}$ has a composite cycle of length k. Hence, $\boldsymbol{A}$ has a composite cycle of length k. Consequently, $k \leqslant c(\boldsymbol{A})$. It follows that $mr(\boldsymbol{A})$ $k \leqslant rank(\boldsymbol{B}) = k \leqslant c(\boldsymbol{A})$.

Shao and Gao[37] showed by concrete examples that the necessary condition in Theorem 2. 2. 2. 7 is not sufficient.

Note that for any sign pattern $\boldsymbol{A}$, $MR(\boldsymbol{A})$ is the maximum number of non-zero entries of $\boldsymbol{A}$ on distinct rows and columns. Thus $c(\boldsymbol{A}) \leqslant MR(\boldsymbol{A})$ holds for every square sign pattern $\boldsymbol{A}$.

Observe that the square of a composite cycle of length k of $\boldsymbol{A} \in Q_n$ can be viewed as a composite cycle of length k of $\boldsymbol{A}^2$, which can be seen from the fact that the square of a generalized permutation sign pattern is again a generalized permutation sign pattern. Hence, for every sign pattern $\boldsymbol{A}$, we have $c(\boldsymbol{A}) \leqslant c(\boldsymbol{A}^2) \leqslant MR(\boldsymbol{A}^2)$. As a consequence, if $mr(\boldsymbol{A}) \leqslant c(\boldsymbol{A})$, then $mr(\boldsymbol{A}) \leqslant c(\boldsymbol{A}) \leqslant c(\boldsymbol{A}^2) \leqslant MR(\boldsymbol{A}^2)$. Therefore, Corollary 2. 2. 2. 5 can be derived from Theorem 2. 2. 2. 7.

The following result is Theorem 2. 5 in [37]. A composite cycle γ of $\boldsymbol{A} \in Q_n$ is said to be chordless if there is no additional arc between the vertices of γ.

Theorem 2. 2. 2. 8 [37] Let $\boldsymbol{A} \in Q_n$. If there exists chordless composite cycle of length k in $\boldsymbol{A}$ with $\mathrm{mr}(\boldsymbol{A}) \leqslant k \leqslant MR(\boldsymbol{A})$, then $\boldsymbol{A}$ allows diagonalizability.

It appears that there is a gap in the proof of this theorem in [37], but the result is valid. Within the proof of this theorem in [37], instead of

$$\boldsymbol{P}^{-1}\boldsymbol{DBP} = \begin{bmatrix} \boldsymbol{C} & \boldsymbol{X} \\ 0 & 0 \end{bmatrix},$$

where $\boldsymbol{C}$ is a non-singular leading principal submatrix of $\boldsymbol{DB}$ with $rank\ \boldsymbol{C} = rank(\boldsymbol{DB})$, we should have

$$\boldsymbol{P}^{-1}\boldsymbol{DBP} = \begin{bmatrix} \boldsymbol{C}' & \boldsymbol{X} \\ 0 & 0 \end{bmatrix},$$

as the similarity via $\boldsymbol{P}$ may cause some changes in the entries of $\boldsymbol{C}$. For example, if

$$\boldsymbol{DB} = \begin{bmatrix} 1 & 2 \\ 2 & 4 \end{bmatrix} \text{ and } \boldsymbol{P}^{-1} = \begin{bmatrix} 1 & 0 \\ 2 & -1 \end{bmatrix}, \text{ then } \boldsymbol{P}^{-1}\boldsymbol{DBP} = \begin{bmatrix} 5 & 2 \\ 0 & 0 \end{bmatrix}.$$

We point out that our proof of Theorem 2. 2. 2. 3 can be adapted to give a proof of this theorem. Indeed, a matrix $\boldsymbol{B} \in Q(\boldsymbol{A})$ of rank k (the existence of such a matrix is guaranteed by Johnson's intermediate rank theorem) is a rank principal matrix due to the existence of a chordless composite cycle of length k (namely, there is a $k \times k$ principal submatrix of $\boldsymbol{A}$ that is a generalized permutation sign pattern).

Theorems 2. 2. 2. 3 and 2. 2. 2. 4 provide useful necessary and sufficient conditions for a sign pattern to allow diagonalizability. However, purely combinatorial characterization of sign patterns that allow diagonalizability is still an open problem that awaits further investigation.

Chapter 3

Rank Conditions for Sign Patterns That Allow Diagonalizability

3.1 Introduction and preliminaries

Sign pattern matrices have been heavily studied and have found applications in other areas such as biology, computer science, neural networks, oriented matroid theory, and convex polytopes theory (see [1, 2, 4, 6, 8, 10, 11, 12, 14, 18, 20, 32, 35, 37, 38]). Sign pattern matrices also form an important part of combinatorial matrix theory, since graph theoretic or combinatorial arguments are often used, as for example in the classical paper[41] by C. Thomassen. Due to the importance of diagonalizable matrices in theory and applications (see [23, 24]), the search for necessary and sufficient conditions characterizing sign patterns that allow diagonalizability has been a long-standing open problem, studied by Eschenbach and Johnson in [9], by Shao and Gao[37,38] and by Feng et al.[14]. The non-negative sign patterns that allow diagonalizability by a unitary matrix, namely, the sign patterns that allow normality, are investigated in [31]. In this chapter, we further investigate sign patterns that allow diagonalizability under matrix rank conditions.

We now introduce some extra definitions and notation, most of which can be found in [10, 14, 20].

Let P be a property referring to a real matrix. For a sign pattern $\boldsymbol{A}$, if there exists a real matrix $\boldsymbol{B} \in Q(\boldsymbol{A})$ such that $\boldsymbol{B}$ has property $\boldsymbol{P}$, we say that $\boldsymbol{A}$ allows $\boldsymbol{P}$; if every $\boldsymbol{B} \in Q(\boldsymbol{A})$ has property $\boldsymbol{P}$, we say that $\boldsymbol{A}$ requires $\boldsymbol{P}$. We focus on sign pattern matrices that allow diagonalizability. We say that a sign pattern $\boldsymbol{A}$ allows diagonalizability with rank k if there is a diagonalizable matrix $\boldsymbol{B} \in Q(\boldsymbol{A})$ with rank k.

Observe that a composite cycle of $\boldsymbol{A}$ corresponds to a disjoint union of cycles in $D(\boldsymbol{A})$, and also corresponds to a non-zero term in the standard expansion of the determinant of a principal submatrix. For two composite cycles γ_1 and γ_2 of $\boldsymbol{A}$, we write $\gamma_1 \subset \gamma_2$ to mean that each simple cycle in γ_1 is also a simple cycle in γ_2 and $\gamma_1 \neq \gamma_2$, or equivalently, if there is a composite cycle β_2 such that $\gamma_2 = \gamma_1 \beta_2$, where the index sets of γ_1 and β_2 are disjoint.

It is well known[20] that $MR(\boldsymbol{A})$ is equal to the maximum number of non-zero entries of $\boldsymbol{A}$ no two of which are on the same line (row or column), which is also equal to the minimum number of lines of $\boldsymbol{A}$ that cover all the non-zero entries of A (see [3]). Hence, for every square sign pattern $\boldsymbol{A}$, we have $c(\boldsymbol{A}) \leqslant MR(\boldsymbol{A})$.

A matrix is said to be full if it does not have any zero entry. The total size of an $m \times n$ matrix is $m+n$. For a matrix $\boldsymbol{A}$, by $\boldsymbol{A}[X, Y]$, we mean the submatrix of $\boldsymbol{A}$ with row index set X and column index set Y. For a matrix $\boldsymbol{A}$, by $\boldsymbol{A}[X]$, we mean the principal submatrix of $\boldsymbol{A}$ whose row index set (as well as the column index set) is X. We say that a zero submatrix of $\boldsymbol{A}$ is maximal if it is not a proper submatrix of any zero submatrix of $\boldsymbol{A}$. We say that certain submatrices of a matrix are strongly disjoint if their row index sets as well as their column index sets are pairwise disjoint.

This chapter is organized as follows. In Section 3.2, we explore necessary and/or sufficient conditions for a square sign pattern in general or of a special type to allow diagonalizability. In Section 3.3, we show that if a square sign pattern $\boldsymbol{A}$ satisfies $mr(\boldsymbol{A}) \leqslant 2$ and $\boldsymbol{A}$ has no zero line, then $\boldsymbol{A}$ allows diagonalizability with rank 2 and also with rank $MR(\boldsymbol{A})$. We also extend some results to sign patterns whose maximal zero submatrices are strongly disjoint. In particular, we establish that for every square sign pattern matrix $\boldsymbol{A}$ whose maximal zero submatrices are strongly disjoint, we have $c(\boldsymbol{A}) = MR(\boldsymbol{A})$ and $\boldsymbol{A}$ has a composite cycle of length $c(\boldsymbol{A})$ consisting of simple cycles of lengths 1 or 2, together with at most one 3-cycle.

A real matrix $\boldsymbol{B}$ is said to be rank-principal if $rank(\boldsymbol{B}) = k$ and $\boldsymbol{B}$ has a nonsingular $k \times k$ principal submatrix $\boldsymbol{C}$. In this book, we call such a principal submatrix $\boldsymbol{C}$ a rank-principal certificate of $\boldsymbol{B}$. Further, we say that a composite cycle γ of a square sign pattern $\boldsymbol{A}$ supports a rank-principal certificate for $\boldsymbol{A}$ if there is a real matrix $\boldsymbol{B} \in Q(\boldsymbol{A})$ that is rank-principal and the index set of γ is equal to the row index set of a rank-principal certificate of $\boldsymbol{B}$. More generally, we say that a composite cycle γ of $\boldsymbol{A}$ supports a principal submatrix $\widehat{\boldsymbol{A}}$ of $\boldsymbol{A}$, if $\widehat{\boldsymbol{A}}$ and γ have the same index set (where the index set of $\widehat{\boldsymbol{A}}$ means its row index set).

For example, it is easily verified that for the sign pattern

$$\begin{bmatrix} 0 & - & - & 0 \\ + & 0 & - & 0 \\ + & - & - & 0 \\ 0 & 0 & + & + \end{bmatrix},$$

the composite cycle $\gamma_1 = (a_{12}a_{21})a_{44}$ supports a rank-principal certificate, but the simple cycle $\gamma_2 = a_{12}a_{23}a_{31}$ does not support a rank-principal certificate.

The following two fundamental sufficient or necessary conditions are contained in [10, 34], using different terminology.

Theorem 3.1.1[10,37] If a square sign pattern $\boldsymbol{A}$ satisfies $c(\boldsymbol{A}) = MR(\boldsymbol{A})$, then $\boldsymbol{A}$ allows diagonalizability with rank $MR(\boldsymbol{A})$.

A square sign pattern that requires non-singularity is said to be sign non-singular. Sign non-singular sign patterns have been studied extensively, such as in the seminal paper[41]. Obviously, every sign non-singular sign pattern $\boldsymbol{A}$ satisfies $c(\boldsymbol{A}) = \mathrm{MR}(\boldsymbol{A})$, and hence, allows diagonalizability.

A special case of Theorem 1 is worth mentioning.

Theorem 3.1.2[37] If a square sign pattern $\boldsymbol{A}$ is combinatorially symmetric, then $c(\boldsymbol{A}) = MR(\boldsymbol{A})$ and $\boldsymbol{A}$ allows diagonalizability with rank $MR(\boldsymbol{A})$.

It is well known that every diagonalizable matrix is rank-principal [14, 24]. It is shown in [14] that every rank-principal matrix is diagonally equivalent to a diagonalizable matrix. The following crucial result, to be used repeatedly in the sequel, was also established in [14].

Theorem 3.1.3[14] A square sign pattern $\boldsymbol{A}$ allows diagonalizability if and only if $\boldsymbol{A}$ allows rank-principality. Furthermore, the sign pattern $\boldsymbol{A}$ allows diagonalizability with rank k if and only if there is a rank-principal matrix $\boldsymbol{B} \in Q(\boldsymbol{A})$ of rank k, or equivalently, if and only if $\boldsymbol{A}$ has a composite cycle of length k that supports a rank-principal certificate for $\boldsymbol{A}$.

By the preceding theorem, the sign patterns that allow diagonalizability are precisely the sign patterns of square rank-principal real matrices. Note that up to permutation similarity, every $n \times n$ rank-principal real matrix of rank $k \geqslant 1$ has the form $\begin{bmatrix} \boldsymbol{B} & \boldsymbol{C} \\ \boldsymbol{D} & \boldsymbol{D}\boldsymbol{B}^{-1}\boldsymbol{C} \end{bmatrix}$, where $\boldsymbol{B}$ is a non-singular matrix of order k while $\boldsymbol{C}$ and $\boldsymbol{D}$ are arbitrary real matrices of suitable sizes. Thus, we can generate many sign patterns that allow diagonalizability quite easily. Even though theoretically every $n \times n$ sign pattern that allows diagonalizability with rank k can arise this way up to permutation similarity, it is difficult to find finitely many choices of $\boldsymbol{B}$, $\boldsymbol{C}$ and $\boldsymbol{D}$ to generate all such sign patterns. Furthermore, we are more interested in combinatorial descriptions of sign patterns that allow diagonalizability.

The following result will be useful in the sequel.

Lemma 3.1.4 Let $\boldsymbol{B}$ be a square real matrix with rank k. Suppose that $\boldsymbol{B}$ has exactly k non-zero eigenvalues. Then B is rank-principal.

Proof Note that $S_k(\boldsymbol{B}) = E_k(\boldsymbol{B})$, where $S_k(\boldsymbol{B})$ is the kth elementary symmetric function of the eigenvalues of $\boldsymbol{B}$ and $E_k(\boldsymbol{B})$ is the sum of all principal minors of order k of $\boldsymbol{B}$ (see [24]). Since $\boldsymbol{B}$ has exactly k non-zero eigenvalues, $S_k(\boldsymbol{B}) \neq 0$. Thus $E_k(\boldsymbol{B}) \neq 0$. It follows that $\boldsymbol{B}$ has at least one non-singular $k \times k$ principal matrix, and hence, $\boldsymbol{B}$ is rank-principal.

For square sign patterns $\boldsymbol{A}$ with minimum rank 1, the necessary condition $c(\boldsymbol{A}) \geqslant mr(\boldsymbol{A})$ is also sufficient for $\boldsymbol{A}$ to allow diagonalizability.

Theorem 3.1.5 Let $\boldsymbol{A}$ be a square sign pattern such that $mr(\boldsymbol{A}) = 1$. Then $\boldsymbol{A}$ allows diagonalizability if and only if $c(\boldsymbol{A}) \geqslant mr(\boldsymbol{A})$. Furthermore, when $c(\boldsymbol{A}) \geqslant mr(\boldsymbol{A})$, for each integer k with $1 \leqslant k \leqslant c(\boldsymbol{A})$, $\boldsymbol{A}$ allows diagonalizability with *rank* k.

Proof The necessity is clear. Suppose that $\boldsymbol{A}$ has a composite cycle γ of length $c(\boldsymbol{A}) \geqslant 1$. Since $mr(\boldsymbol{A}) = 1$, any two non-zero rows of $\boldsymbol{A}$ are either identical or are negatives of each other, so the principal submatrix of $\boldsymbol{A}$ supported by γ has no zero entries. Let $\boldsymbol{B}$ be the $(1, -1, 0)$ -matrix in $Q(\boldsymbol{A})$. Clearly, $\text{rank}(\boldsymbol{B}) = 1$, $\boldsymbol{B}$ is rank-principal, and the principal submatrix supported by γ has all the diagonal entries non-zero. For each integer k with $1 \leqslant k \leqslant c(\boldsymbol{A})$, by doubling $k-1$ diagonal entries of the principal submatrix of $\boldsymbol{B}$ supported by γ, we obtain a rank-principal matrix of rank k in $Q(\boldsymbol{A})$, which ensures that $Q(\boldsymbol{A})$ contains a diagonalizable matrix with rank k by the preceding theorem.

For a sign pattern that allows diagonalizability, it is natural to consider the possible ranks of the diagonalizable matrices in the qualitative class. By Theorem 3.1.3, a sign pattern $\boldsymbol{A}$ allows diagonalizability with rank k (of course, $k \geqslant mr(\boldsymbol{A})$) if and only if $\boldsymbol{A}$ has a composite cycle of length k that supports a rank-principal certificate for $\boldsymbol{A}$. A natural question is: for a sign pattern $\boldsymbol{A}$ that allows diagonalizability, can every composite cycle γ of $\boldsymbol{A}$ with length at least $mr(\boldsymbol{A})$ support a rank-principal certificate? The answer is negative, as the following two examples show.

Example 3.1.6 Consider the reducible sign pattern $A = \begin{bmatrix} + & + & 0 \\ + & + & 0 \\ 0 & 0 & + \end{bmatrix}$.

Note that $mr(A) = 2$, $MR(A) = 3$. The maximum length composite cycle $\gamma_1 = a_{11} a_{22} a_{33}$ supports a rank-principal certificate. The composite cycle $\gamma_2 = a_{22} a_{33}$ supports a rank-principal certificate of order 2. But the composite cycle $\gamma_3 = a_{11} a_{22}$ cannot support a rank-principal certificate, since the third row of any matrix in $Q(A)$ cannot be in the span of the first two rows.

Example 3.1.7 Consider the irreducible sign pattern

$$A = \begin{bmatrix} 0 & + & + & 0 & + \\ 0 & + & + & 0 & + \\ 0 & 0 & 0 & + & 0 \\ 0 & + & + & 0 & + \\ + & 0 & 0 & 0 & 0 \end{bmatrix}$$

Note that $mr(A) = 3$ and $c(A) = MR(A) = 5$, so A allows diagonalizability with rank 5. Observe that A has several composite cycles of length 3 (such as $a_{23} a_{34} a_{34}$), but no composite cycle of length 3 can support a rank-principal certificate. Indeed, for every matrix in $Q(A)$, the third and fifth rows are not in the span of the other rows, so every composite cycle that supports a rank-principal certificate must contain the indices 3 and 5; similarly, by examining the columns 1 and 4, we see that every composite cycle that supports a rank-principal certificate must contain the indices 1 and 4. Thus every composite cycle that supports a rank-principal certificate must contain the indices 1, 3, 4, and 5, and hence must have length at least 4. Therefore, A does not allow diagonalizability with rank $mr(A) = 3$.

As illustrated in the preceding example, a sign pattern A that allows diagonalizability may not allow diagonalizability with rank $mr(A)$, even when $mr(A)$ is the length of a composite cycle.

But for a bipartite sign pattern (a sign pattern whose digraph is bipartite), we have the following interesting result.

Theorem 3.1.8[19] Let A be a symmetric bipartite sign pattern. Then $mr(A)$, $MR(A)$ and the length of every composite cycle of A are even, and for every even integer k with $mr(A) \leqslant k \leqslant MR(A)$, there is a symmetric (and hence diagonalizable) matrix $\boldsymbol{B} \in Q(\boldsymbol{A})$ with *rank* k.

It follows from the preceding theorem that for any symmetric bipartite sign pattern $\boldsymbol{A}$ and every even integer k with $mr(\boldsymbol{A}) \leqslant k \leqslant MR(\boldsymbol{A})$, there is a composite cycle of $\boldsymbol{A}$ of length k that supports a rank-principal certificate for $\boldsymbol{A}$. However, even for a symmetric irreducible bipartite sign pattern $\boldsymbol{A}$, not every composite cycle of $\boldsymbol{A}$ of length at least $\mathrm{mr}(\boldsymbol{A})$ can support a rank-principal certificate for $\boldsymbol{A}$, as the following example shows.

Example 3.1.9 Consider the symmetric irreducible bipartite sign pattern

$$\boldsymbol{A} = \begin{bmatrix} 0 & 0 & 0 & + & + & 0 \\ 0 & 0 & 0 & + & + & 0 \\ 0 & 0 & 0 & + & + & + \\ + & + & + & 0 & 0 & 0 \\ + & + & + & 0 & 0 & 0 \\ 0 & 0 & + & 0 & 0 & 0 \end{bmatrix}$$

Clearly, $mr(\boldsymbol{A}) = 4$. But for every real matrix $\boldsymbol{B} \in Q(\boldsymbol{A})$ with rank 4, the first two rows as well as the first two columns must be linearly dependent. Thus the composite cycle $(a_{14}a_{41})(a_{25}a_{52})$ cannot support a rank-principal certificate of $\boldsymbol{A}$.

Hall et al. [19] also shows that for some symmetric sign patterns $\boldsymbol{A}$, $mr(\boldsymbol{A})$ cannot be achieved by any symmetric matrix $\boldsymbol{B} \in Q(\boldsymbol{A})$. The following two natural questions arise.

Problem 3.1.10 Does every symmetric sign pattern $\boldsymbol{A}$ allow diagonalizability with rank $mr(\boldsymbol{A})$?

Problem 3.1.11 Is it true that for every irreducible symmetric sign pattern $\boldsymbol{A}$ and every integer k that is the length of some composite cycle of $\boldsymbol{A}$ with $k \geqslant mr(\boldsymbol{A})$, there is a composite cycle of $\boldsymbol{A}$ of length k that supports a rank-principal certificate for $\boldsymbol{A}$?

We point out that if symmetry is related to combinatorial symmetry, the answers to the two preceding problems are negative, as the following example shows.

Example 3.1.12 Consider the combinatorially symmetric irreducible bipartite sign pattern

$$A=\begin{bmatrix} 0 & 0 & 0 & + & + & + & + \\ 0 & 0 & 0 & + & + & + & + \\ 0 & 0 & 0 & + & + & + & + \\ + & + & + & 0 & 0 & 0 & 0 \\ - & + & + & 0 & 0 & 0 & 0 \\ + & - & + & 0 & 0 & 0 & 0 \\ + & + & - & 0 & 0 & 0 & 0 \end{bmatrix},$$

Observe that the 4×3 submatrix in the lower left corner has minimum rank 3 (see [4]), so $mr(\boldsymbol{A})=4$. Assume that a rank 4 matrix $\boldsymbol{B}\in Q(\boldsymbol{A})$ is rank-principal. Then there is a composite cycle γ of length 4 that supports a rank-principal certificate of $\boldsymbol{B}$. Since the first three columns of $\boldsymbol{B}$ are linearly independent and are not linear combinations of the remaining columns, we see that the index set of γ must contain $\{1, 2, 3\}$. It follows that the principal submatrix supported by γ would have row indices $\{1, 2, 3, i\}$ for some $i\in\{4, 5, 6, 7\}$. Thus the principal submatrix of $\boldsymbol{B}$ supported by γ, $\boldsymbol{B}[\{1, 2, 3, i\}]$, contains a 3×3 zero submatrix and has rank 2, contradicting the fact that it is a rank-principal certificate of $\boldsymbol{B}$. Thus $\boldsymbol{A}$ does not allow diagonalizability with rank 4. Therefore, no composite cycle of length 4 can support a rank-principal certificate for $\boldsymbol{A}$. Note, however, that $\boldsymbol{A}$ does have composite cycles of length 4, such as $(a_{34}a_{43})(a_{25}a_{52})$.

It is easy to see that the answer to Problem 3. 1. 11 is negative if irreducibility is dropped, as can be seen from the following example.

Example 3. 1. 13 For the reducible symmetric sign pattern

$$A=\begin{bmatrix} 0 & + & + & 0 & 0 & 0 & 0 \\ + & 0 & + & 0 & 0 & 0 & 0 \\ + & + & 0 & 0 & 0 & 0 & 0 \\ 0 & 0 & 0 & 0 & 0 & + & + \\ 0 & 0 & 0 & 0 & 0 & + & + \\ 0 & 0 & 0 & + & + & 0 & 0 \\ 0 & 0 & 0 & + & + & 0 & 0 \end{bmatrix},$$

$mr(\boldsymbol{A})=5$, and there exist composite cycles of length 6 (such as $(a_{12}a_{21})(a_{46}a_{64})(a_{57}a_{75})$ Clearly, every composite cycle that supports a rank-principal certificate for $\boldsymbol{A}$ must contain the indices 1, 2 and 3. But there is no composite cycle of length 6 containing the indices 1, 2 and 3.

Concerning composite cycles that support rank-principal certificates, we have the following result, which will be used in the next section.

Theorem 3.1.14 Suppose that γ_1 and γ_2 are composite cycles of a square sign pattern $\boldsymbol{A}$ such that $\gamma_1 \subset \gamma_2$. If γ_1 supports a rank-principal certificate for $\boldsymbol{A}$, then γ_2 also supports a rank-principal certificate for $\boldsymbol{A}$.

Proof Without loss of generality, we may assume that the sign pattern $\boldsymbol{A}$ has the following form $\boldsymbol{A} = \begin{bmatrix} \boldsymbol{A}_{11} & \boldsymbol{A}_{12} & \boldsymbol{A}_{13} \\ \boldsymbol{A}_{21} & \boldsymbol{A}_{22} & \boldsymbol{A}_{23} \\ \boldsymbol{A}_{31} & \boldsymbol{A}_{32} & \boldsymbol{A}_{33} \end{bmatrix}$, where $\boldsymbol{A}_{11}$ is supported by γ_1 and $\begin{bmatrix} \boldsymbol{A}_{11} & \boldsymbol{A}_{12} \\ \boldsymbol{A}_{21} & \boldsymbol{A}_{22} \end{bmatrix}$ is supported by γ_2. Since $\gamma_1 \subset \gamma_2$, there is a composite cycle β_2 of $\boldsymbol{A}$ such that $\gamma_2 = \gamma_1 \beta_2$, where the index sets of γ_1 and β_2 are disjoint. It follows that $\boldsymbol{A}_{22}$ is supported by β_2. Suppose that γ_1 supports a rank-principal certificate of $\boldsymbol{B} = \begin{bmatrix} \boldsymbol{B}_{11} & \boldsymbol{B}_{12} & \boldsymbol{B}_{13} \\ \boldsymbol{B}_{21} & \boldsymbol{B}_{22} & \boldsymbol{B}_{23} \\ \boldsymbol{B}_{31} & \boldsymbol{B}_{32} & \boldsymbol{B}_{33} \end{bmatrix} \in Q(\boldsymbol{A})$. Let k_1 and k_2 denote the lengths of γ_1 and β_2, respectively. By performing type Ⅲ elementary row and column operations on $\boldsymbol{B}$, we can get the following matrix of rank k_1:

$$\begin{bmatrix} \boldsymbol{B}_{11} & 0 & 0 \\ 0 & 0 & 0 \\ 0 & 0 & 0 \end{bmatrix}.$$

Let $\boldsymbol{B}_{22}(\beta_2)$ be the $(1, -1, 0)$ -matrix of order k_2 whose only non-zero entries occur in the positions of the entries of β_2 and have the same sign as the corresponding entries of β_2. Clearly, $\boldsymbol{B}_{22}(\beta_2)$ is non-singular. Let $\tilde{\boldsymbol{B}}$ be the matrix obtained from $\boldsymbol{B}$ by replacing $\boldsymbol{B}_{22}$ with $\boldsymbol{B}_{22} + \boldsymbol{B}_{22}(\beta_2)$ while keeping the other blocks unchanged. Note that $\tilde{\boldsymbol{B}} \in Q(\boldsymbol{A})$. It can be seen that via type Ⅲ elementary row and column operations, $\tilde{\boldsymbol{B}}$ may be transformed to the following matrix:

$$\begin{bmatrix} \boldsymbol{B}_{11} & 0 & 0 \\ 0 & \boldsymbol{B}_{22}(\beta_2) & 0 \\ 0 & 0 & 0 \end{bmatrix}.$$

It follows that $\tilde{B}$ is rank-principal, with a rank-principal certificate supported by γ_2.

However, the following problem remains open.

Problem 3.1.15 Suppose that γ_1 and γ_2 are composite cycles of a square sign pattern A such that γ_1 supports a rank-principal certificate for A, and the index set of γ_1 is a subset of that of γ_2. Does it follow that γ_2 also supports a rank-principal certificate for A?

It can be seen from Theorem 3.1.14 that for every $n \times n$ sign pattern with all diagonal entries non-zero, if A allows diagonalizability with *rank* k, then it also allows diagonalizability with rank t for each integer t with $k \leqslant t \leqslant n$. The following intriguing question arises.

Problem 3.1.16 Does every square sign pattern A with all diagonal entries non-zero allow diagonalizability with rank $mr(A)$?

3.2 Main results

It is shown in [38] that for each integer $k \geqslant 4$, there exists an irreducible square sign pattern A such that $c(A) \geqslant mr(A) = k$ and A does not allow diagonalizability. For instance, the following is such an irreducible sign pattern A with $mr(A) = 4$.

Example 3.2.1[38] Let

$$A = \begin{bmatrix} 0 & + & + & 0 & + & 0 \\ 0 & + & + & 0 & + & 0 \\ 0 & 0 & 0 & + & 0 & 0 \\ + & 0 & 0 & 0 & 0 & 0 \\ 0 & 0 & 0 & 0 & 0 & + \\ + & 0 & 0 & 0 & 0 & 0 \end{bmatrix}.$$

Figure 1. Digraph of A in Example 2.2.1

It can be seen that A is irreducible (by checking that the digraph $D(A)$ shown in Figure 1 is strongly connected), $mr(A) = c(A) = 4$, and $MR(A) = 5$. As shown in [39], A does not allow diagonalizability. We give a simpler argument using rank-principality to show that A does not allow diagonalizability. Note that the entries $a_{34} = +$ and $a_{56} = +$ are the only non-zero entries in their rows and columns, so if there is a composite cycle γ that supports a rank-principal certificate for A, then γ must contain the entries $a_{34}=+$ and $a_{56}=+$, and hence the index set of γ would contain $\{3, 4, 5, 6\}$. But the digraph $D(A)$ (See Figure 1) shows that there is no composite cycle in A whose index set contains $\{3, 4, 5, 6\}$. Thus A does not allow diagonalizability.

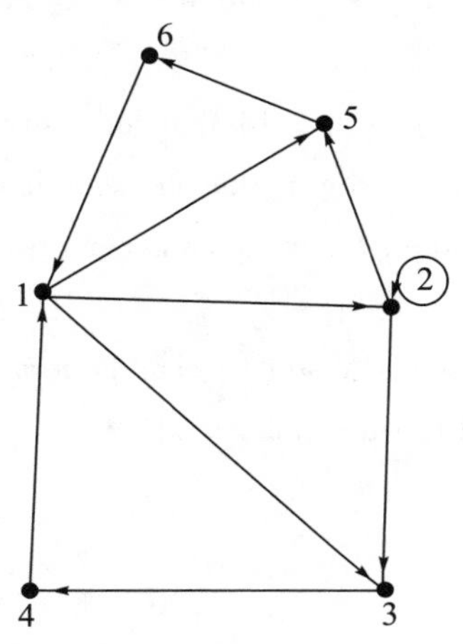

We point out a way to construct irreducible sign patterns A such that $mr(A)=2+t\geqslant 4$ and $c(A)\leqslant 3$ (and hence A does not allow diagonalizability). For each integer $t\geqslant 2$, the non-negative sign pattern $A=[a_{ij}]$ of order $2t+1$ whose only non-zero entries are the entries of the 3-cycles $a_{1,2i}a_{2i,2i+1}a_{2i+1,1}$, $i=1, 2, \cdots, t$ satisfies $mr(A)=2+t\geqslant 4$ and $c(A)=3$. In contrast, the following remarkable property of irreducible sign patterns with minimum rank 3 is worth mentioning.

Observation 3.2.2 Every irreducible sign pattern A with $mr(A)=3$ satisfies $c(A)\geqslant 3$.

Proof If A has a simple cycle of length at least 3, then of course $c(A)\geqslant 3$. Now assume that the maximum simple cycle length of A is 2. Since $D(A)$ is strongly connected, we see that A is combinatorially symmetric and the underlying undirected graph $G(A)$ of $D(A)$ is a tree. If $G(A)$ is not a star, then $G(A)$ has two non-adjacent edges, and hence, we get a composite cycle of A of length 4 consisting of two simple 2-cycles. If $G(A)$ is a star, then $mr(A)=3$ ensures that there is a 1-cycle at a vertex that is not the center of the star, thus we get a composite cycle of length 3 consisting of this 1-cycle and a 2-cycle.

Thus we have the following problem.

Problem 3.2.3 Does every irreducible sign pattern $\boldsymbol{A}$ with $mr(\boldsymbol{A}) = 3$ allow diagonalizability?

This is a very challenging problem that awaits further research. The irreducible sign patterns with minimum rank 3 that we have examined so far all allow diagonalizability. But note that as shown in Example 3.1.7, irreducible sign patterns with minimum rank 3 may not allow diagonalizability with rank 3.

If we relax the irreducibility to the weaker condition of having no zero line, then it is easy to find a square sign pattern with minimum rank 3 and with no zero line that does not allow diagonalizability, as the next example shows.

Example 3.2.4 The reducible sign pattern $\boldsymbol{A} = \begin{bmatrix} + & + & 0 & 0 \\ + & + & 0 & 0 \\ 0 & 0 & 0 & + \\ 0 & 0 & 0 & 0 \end{bmatrix}$ has no zero line, $mr(\boldsymbol{A}) = 3$, and $c(\boldsymbol{A}) = 2$. By Theorem 2.2.2.7, $\boldsymbol{A}$ does not allow diagonalizability.

We now concentrate on sign patterns with minimum rank 2. If the irreducibility condition is dropped completely, it is easy to see that there exists a reducible sign pattern $\boldsymbol{A}$ such that $mr(\boldsymbol{A}) = 2$ and $\boldsymbol{A}$ does not allow diagonalizability. For example, the sign pattern

$$\boldsymbol{A} = \begin{bmatrix} + & + & 0 & 0 \\ + & + & 0 & 0 \\ 0 & 0 & 0 & + \\ 0 & 0 & 0 & 0 \end{bmatrix}$$

satisfies $c(\boldsymbol{A}) = mr(\boldsymbol{A}) = 2$, but $\boldsymbol{A}$ does not allow diagonalizability (as its lower right 2×2 diagonal block does not allow diagonalizability).

In the following theorems, we establish that every square sign pattern $\boldsymbol{A}$ satisfying the two conditions $mr(\boldsymbol{A}) = 2$ and $\boldsymbol{A}$ has no zero line (where the second condition is weaker than irreducibility) allows diagonalizability with various ranks including 2 and $MR(\boldsymbol{A})$.

Theorem 3.2.5 Let $\boldsymbol{A}$ be a square sign pattern such that $mr(\boldsymbol{A}) = 2$ and $\boldsymbol{A}$ has no zero line. Then $\boldsymbol{A}$ allows diagonalizability with rank 2.

Proof Since the set of sign patterns that allow diagonalizability is closed under permutation similarity and signature equivalence, using a method similar to the proof of Theorem 2.6 of [12], by replacing the sign pattern $\boldsymbol{A}$ with a sign pattern obtained from $\boldsymbol{A}$ via permutation similarity and signature equivalence if necessary, we may assume that a real matrix $\boldsymbol{B} \in Q(\boldsymbol{A})$ with $rank(\boldsymbol{B}) = 2$ can be written as

$$B = \begin{bmatrix} 1 & x_1 \\ 1 & x_2 \\ \vdots & \vdots \\ 1 & x_n \end{bmatrix} \begin{bmatrix} -y_1z_1 & -y_2z_2 & \cdots & -y_nz_n \\ z_1 & z_2 & \cdots & z_n \end{bmatrix} = ((x_i - y_j)\ z_j),$$

where $x_1 \leqslant x_2 \leqslant \cdots \leqslant x_n$, $z_1 > 0$, $z_2 > 0$, $\cdots$, $z_n > 0$, not all of the x_i are equal, and not all the y_i are equal. Regard x_1, x_2, $\cdots$, x_n, y_1, y_2, $\cdots$, y_n and z_1, z_2, $\cdots$, z_n as real variables. For each i, j, whenever $b_{ij} = 0$, we identify y_j with x_i. Furthermore, whenever $b_{i_1j} = b_{i_2j} = 0$, we also identify x_{i_1} with x_{i_2}. Consider the following 2×2 matrix (obtained by switching the two factors in the above full rank factorization of B)

$$C = \begin{bmatrix} -y_1z_1 & -y_2z_2 & \cdots & -y_nz_n \\ z_1 & z_2 & \cdots & z_n \end{bmatrix} \begin{bmatrix} 1 & x_1 \\ 1 & x_2 \\ \vdots & \vdots \\ 1 & x_n \end{bmatrix} = \begin{bmatrix} -\sum_{i=1}^{n} y_iz_i & -\sum_{i=1}^{n} x_iy_iz_i \\ \sum_{i=1}^{n} z_i & -\sum_{i=1}^{n} x_iz_i \end{bmatrix},$$

whose determinant is the polynomial $p = (\sum_{i=1}^{n} z_i)(\sum_{i=1}^{n} x_iy_iz_i) - (\sum_{i=1}^{n} x_iz_i)(\sum_{i=1}^{n} y_iz_i)$.

Observe that if $\boldsymbol{C}$ is non-singular, then $\boldsymbol{C}$ has two non-zero eigenvalues, hence $\boldsymbol{B}$ has exactly two non-zero eigenvalues. In this case, by Lemma 3.1.4, $\boldsymbol{B}$ is rank-principal, and hence, by Theorem 3.1.3, $\boldsymbol{A}$ allows diagonalizability. Thus it remains to show that the independent variables can be assigned suitable values so that $p \neq 0$, which ensures that $\boldsymbol{C}$ is non-singular.

Since $rank(\boldsymbol{B}) = 2 > 1$, at least two of the x_i are independent, and at least two of the y_i are independent. Suppose that there is no zero entry in row 1 of $\boldsymbol{B}$. Then x_1 is distinct from x_2, x_3, $\cdots$, x_n and no y_j is identified as x_1. Hence, if y_1 is not identified with any x_i, then the coefficient of x_1y_1 in the polynomial p is $(\sum_{i=1}^{n} z_i)z_1$

$-z_1^2 \neq 0$, so $p \neq 0$. If $y_1 = x_{i_1} = \cdots = x_{i_k} = y_{j_2} = \cdots = y_{j_t}$, then the coefficient of $x_1 x_{i_1}$ in p is $(\sum_{i=1}^{n} z_i) z_1 - z_1 (\sum_{j=1}^{k} z_{i_j}) \neq 0$.

Now assume that row 1 of $\boldsymbol{B}$ has exactly $t(>0)$ zero entries, with column indices $j_1 < j_2 < \cdots < j_t$. Suppose that the j_1 th column of $\boldsymbol{B}$ has k zero entries. Since each column of $\boldsymbol{B}$ is nondecreasing (as the row indices increase), we have $b_{1j_1} = b_{2j_1} = \cdots = b_{kj_1} = 0$ and the remaining entries in this column are positive. Because $mr(\boldsymbol{A}) = 2 = rank(\boldsymbol{B})$ and $\boldsymbol{A}$ has no zero line, we have $1 \leqslant k < n$ and $1 \leqslant t < n$ and $\boldsymbol{B}[\{1, 2, \cdots, k\}, \{j_1, j_2, \cdots, j_t\}]$ is a maximal zero submatrix such that all entries of $\boldsymbol{B}[\{1, 2, \cdots, k\}, \{j_1, j_2, \cdots, j_t\}^c]$ are non-zero and all the entries of $\boldsymbol{B}[\{1, 2, \cdots, k\}^c, \{j_1, j_2, \cdots, j_t\}]$ are non-zero. Suppose that there are s elements in $\{j_1, j_2, \cdots, j_t\}$ that are at most k, that is, $\boldsymbol{B}[\{1, 2, \cdots, k\}, \{j_1, j_2, \cdots, j_t\}]$ contains s diagonal entries of $\boldsymbol{B}$. Note that $x_1 = x_2 = \cdots = x_k = y_{j_1} = y_{j_2} = \cdots = y_{j_t}$, further $x_{k+1}, \cdots, x_n$ and y_j for any $j \notin \{j_1, j_2, \cdots, j_t\}$ are independent of x_1. Thus the coefficient of x_1^2 in p is

$$(\sum_{i=1}^{n} z_i)(\sum_{i=1}^{s} z_{j_i}) - (\sum_{i=1}^{k} z_i)(\sum_{i=1}^{t} z_{j_i}),$$

which is not identically equal to 0, since $1 \leqslant k < n$, $1 \leqslant t < n$, and the polynomial ring $\mathbb{R}[z_1, z_2, \cdots, z_n]$ is a unique factorization domain. (Note that $\sum_{i=1}^{s} z_{j_i}$ is understood to be 0 when $s = 0$.)

Therefore, the polynomial p is never identically zero. Thus, subject to the required identifications, we can find a rational value for each free variable within a sufficiently small neighborhood of its initial value such that $p \neq 0$ at the perturbed point, and the perturbed rational matrix $\tilde{\boldsymbol{B}}$ given by the first displayed factorization in this proof using the perturbed values of the variables satisfies $\tilde{\boldsymbol{B}} \in Q(\boldsymbol{A})$ (as the non-zero entries keep the same sign due to continuity, while the zero entries are preserved due to the variable identifications), $rank(\tilde{\boldsymbol{B}}) = 2$, and $\tilde{\boldsymbol{B}}$ has two non-zero eigenvalues. By Lemma 3.1.4, $\tilde{\boldsymbol{B}}$ is rank-principal. It follows that $\boldsymbol{A}$ allows diagonalizability with rank 2.

As an immediate consequence, we get the following result.

Theorem 3.2.6 Every irreducible sign pattern $\boldsymbol{A}$ with $mr(\boldsymbol{A}) = 2$ allows diagonalizability with rank 2.

Note that in the proof of Theorem 3.2.5, if $\boldsymbol{A}[X, Y]$ is a maximal zero submatrix of the square sign pattern A such that $mr(\boldsymbol{A}) = 2$ and $\boldsymbol{A}$ has no zero line, then $\boldsymbol{A}[X, Y^c]$ and $\boldsymbol{A}[X^c, Y]$ are full sign patterns. Hence, for any other maximal zero submatrix $\boldsymbol{A}[X_1, Y_1]$ of $\boldsymbol{A}$, we must have $X \cap X_1 = \varphi$ and $Y \cap Y_1 = \varphi$, Thus the maximal zero submatrices of $\boldsymbol{A}$ are strongly disjoint. It is easy to see that this holds even when $\boldsymbol{A}$ is not square. We record this fact as follows.

Observation 3.2.7 Let $\boldsymbol{A}$ be a sign pattern such that $mr(\boldsymbol{A}) = 2$ and $\boldsymbol{A}$ has no zero line. Then the maximal zero submatrices of $\boldsymbol{A}$ are strongly disjoint.

It turns out every square sign pattern whose maximal zero submatrices (if any) are strongly disjoint allows diagonalizability with rank equal to its maximum rank, as shown below.

Theorem 3.2.8 Let $\boldsymbol{A}$ be an $n \times n$ sign pattern whose maximal zero submatrices (if any) are strongly disjoint. Then $c(\boldsymbol{A}) = MR(\boldsymbol{A})$ and $\boldsymbol{A}$ allows diagonalizability with rank $MR(\boldsymbol{A})$. Furthermore, for every $n \times n$ permutation sign pattern $\boldsymbol{P}$, $c(\boldsymbol{PA}) = MR(\boldsymbol{PA}) = MR(\boldsymbol{AP}) = c(\boldsymbol{AP}) = MR(\boldsymbol{A})$.

Proof In view of Theorem 3.1.1, it suffices to show that $c(\boldsymbol{A}) = MR(\boldsymbol{A})$, as the last statement of the theorem follows from this fact applied to the matrices $\boldsymbol{PA}$ and $\boldsymbol{AP}$ (and the obvious fact that the maximum rank is invariant under permutation equivalence). Clearly, $c(\boldsymbol{A}) = MR(\boldsymbol{A})$ when $MR(\boldsymbol{A}) = n$.

Now assume that $MR(\boldsymbol{A}) < n$. Then there are s rows, with row index set S, and t columns, with column index set T, that cover all the non-zero entries of A, where $s + t = |S| + |T| = MR(\boldsymbol{A})$. Then $\boldsymbol{A}[S^c, T^c] = 0$ is a maximal zero submatrix. Since the maximal zero submatrices of $\boldsymbol{A}$ are strongly disjoint, $\boldsymbol{A}[S, T^c]$ and $\boldsymbol{A}[S^c, T]$ are full. Let $k = |S \cap T|$. Note that each element $z \in (S \setminus T) \cup (T \setminus S)$ gives a 1-cycle a_{zz}, so we have $|S| + |T| - 2k$ disjoint 1-cycles of $\boldsymbol{A}$ arising this way.

For each $y \in (S \cup T)^c$ and $x \in S \cap T$, we have $a_{xy} \neq 0$ and $a_{yx} \neq 0$, so $a_{xy}a_{yx}$ is a 2-cycle. Since $|(S \cup T)^c| = n - (|S| + |T| - k) = k + n - (|S| + |T|) \geqslant k$, we obtain k disjoint 2-cycles using k disjoint pairs of vertices $x_i \in S \cap T$ and $y_i \in (S \cup T)^c$, $i = 1, 2, \cdots, k$. Together with the $|S| + |T| - 2k$ disjoint 1-cycles mentioned above, we obtain a composite cycle of length $2k + (|S| + |T| - 2k) = |S| + |T| = MR(\boldsymbol{A})$ Thus $c(\boldsymbol{A}) = MR(\boldsymbol{A})$.

As a consequence of Theorem 3. 2. 8 and Observation 3. 2. 7, we get the following result.

Theorem 3. 2. 9 Let $\boldsymbol{A}$ be an $n \times n$ sign pattern such that $mr(\boldsymbol{A}) = 2$ and $\boldsymbol{A}$ has no zero line. Then $c(\boldsymbol{A}) = MR(\boldsymbol{A})$ and $\boldsymbol{A}$ allows diagonalizability with rank $MR(\boldsymbol{A})$. Furthermore, for every $n \times n$ permutation sign pattern $\boldsymbol{P}$, $c(\boldsymbol{PA}) = MR(\boldsymbol{PA}) = MR(\boldsymbol{AP}) = c(\boldsymbol{AP}) = MR(\boldsymbol{A})$.

In the proof of Theorem 3. 2. 5, the presence of zero entries in $\boldsymbol{A}$ imposes restrictions on some of the variables in a full rank factorization of a matrix $\boldsymbol{B} \in Q(\boldsymbol{A})$. However, for any $n \times n$ full sign pattern $\boldsymbol{A}$ and any $\boldsymbol{B} \in Q(\boldsymbol{A})$ with $rank(\boldsymbol{B}) = mr(\boldsymbol{A})$, there are no such restrictions on the variables arising from a full rank factorization of $\boldsymbol{B}$, namely, all the variables involved are independent variables. Hence, using a full rank factorization as in the proof of Theorem 3. 2. 5, we can show that $\boldsymbol{A}$ allows diagonalizability with rank $mr(\boldsymbol{A})$. Combined with Theorem 3. 1. 14, we obtain the following result.

Theorem 3. 2. 10 Every $n \times n$ full sign pattern $\boldsymbol{A}$ allows diagonalizability with each rank from $mr(\boldsymbol{A})$ to n.

Sign patterns whose maximal zero submatrices are strongly disjoint may be viewed as a generalization of full sign patterns, but it could happen that such a square sign pattern $\boldsymbol{A}$ may not allow diagonalizability with any rank less than its maximum rank, as the following example shows.

Example 3. 2. 11 The maximal zero submatrices of the square sign pattern $\boldsymbol{A} = \begin{bmatrix} 0 & 0 & + & + \\ 0 & 0 & + & - \\ + & + & 0 & 0 \\ + & + & 0 & 0 \end{bmatrix}$ are strongly disjoint, and $\boldsymbol{A}$ allows diagonalizability with rank $c(\boldsymbol{A}) = 4$. But $mr(\boldsymbol{A}) = 3$ and $\boldsymbol{A}$ does not have any composite cycle of length 3 (as $D(\boldsymbol{A})$ is bipartite), so $\boldsymbol{A}$ does not allow diagonalizability with rank 3.

We now show another striking composite cycle property of square sign patterns whose maximal zero submatrices are strongly disjoint.

Theorem 3. 2. 12 Let $\boldsymbol{A}$ be an $n \times n$ non-zero sign pattern whose maximal zero submatrices (if any) are strongly disjoint. Then $\boldsymbol{A}$ has a composite cycle of length $c(\boldsymbol{A})$ consisting of disjoint simple cycles of lengths up to 3, at most one of which is a 3-cycle.

Proof We proceed by induction on n.

The result is clear for $n \leqslant 3$.

Note that for $n=3$, such as for the sign pattern $\begin{bmatrix} 0 & - & - \\ + & 0 & - \\ + & + & 0 \end{bmatrix}$, it is possible that the only composite cycle of length 3 is a simple 3-cycle.

Now, assume that $n \geqslant 4$ and suppose that the result holds for all orders less than n.

If $\boldsymbol{A}$ has no zero submatrices, then clearly $\boldsymbol{A}$ has a composite cycle of length n consisting of n 1-cycles.

Now assume that $\boldsymbol{A}$ has $m \geqslant 1$ strongly disjoint maximal zero submatrices and without loss of generality, suppose that the row index sets of the maximal zero submatrices of $\boldsymbol{A}$ are the pairwise disjoint subsets $S_1, S_2, \cdots, S_m$, and their column index sets are the pairwise disjoint subsets $T_1, T_2, \cdots, T_m$, where $|S_1|+|T_1| \geqslant |S_2|+|T_2| \geqslant \cdots \geqslant |S_m|+|T_m|$.

Case 1. $|S_1|+|T_1|>n$.

Then fewer than n lines of $\boldsymbol{A}$ (such as rows and columns of $\boldsymbol{A}$ not intersecting $\boldsymbol{A}[S_1, T_1]$) can cover all the non-zero entries of A, so $MR(\boldsymbol{A})<n$. As in the proof of Theorem 3.2.8 when $MR(\boldsymbol{A})<n$, there is a composite cycle of length $c(\boldsymbol{A})$ consisting of 1-cycles and 2-cycles.

Case 2. $|S_1|+|T_1|=n$.

Since the total size of any zero submatrix of $\boldsymbol{A}$ is at most n, we have $MR(\boldsymbol{A})=n$. Clearly, $S_1^c \neq \varphi$ and $T_1^c \neq \varphi$. Note that every zero submatrix of $\boldsymbol{A}$ strongly disjoint with $\boldsymbol{A}[S_1, T_1]$ is a submatrix of $\boldsymbol{A}[S_1^c, T_1^c]$. By avoiding using any possible non-zero entries in $\boldsymbol{A}[S_1^c, T_1^c]$ in forming a composite cycle of length n, we may assume that $S_2=S_1^c$ and $T_2=T_1^c$ (and hence $m=2$). Note that we then have $|S_2|+|T_2|=n-|S_1|+n-|T_1|=2n-(|S_1|+|T_1|)=n$.

Subcase 2.1. $S_1=T_1$.

Take $i \in S_1=T_1$, $j \in S_2=T_2$. Then $a_{ij}a_{ji}$ is a 2-cycle in $\boldsymbol{A}$. Upon deleting i th and j th rows and columns of $\boldsymbol{A}$, each of the two maximal zero submatrices of $\boldsymbol{A}$ loses one row and one column, and the principal submatrix $A' = A[\{i, j\}^c]$ is of order $n-2$ and $MR(A') = n-2$, as every zero submatrix of A' has total size at most $n-2$. By the induction hypothesis, A' has composite cycle γ of length $c(A') = n-2$ consisting of 1-cycles, 2-cycles, and at most one 3-cycle. Thus $a_{ij}a_{ji}\gamma$ is a composite cycle of A of length $n = c(A)$ consisting of 1-cycles, 2-cycles, and at most one 3-cycle.

Subcase 2.2. $S_1 \neq T_1$.

Then $(S_1 \setminus T_1) \cup (T_1 \setminus S_1) \neq \varphi$. Without loss of generality, assume that $(S_1 \setminus T_1) \neq \varphi$ and take $k \in S_1 \setminus T_1$. Then $a_{kk} \neq 0$, as it is an element of $A[S_1, T_1^c] = A[S_1, T_2]$. Upon deleting the k th row and k th column of A, each of the two maximal zero submatrices of A loses one line, and we get a principal submatrix A of order $n-1$ with $MR(A') = n-1$, since A' does not have any zero submatrix of total size greater than $n-1$. By the induction hypothesis, A' has a composite cycle γ of length $c(A') = n-1$ consisting of 1-cycles, 2-cycles, and at most one 3-cycle. Thus $a_{kk}\gamma$ is a composite cycle of A of length $n = c(A)$ consisting of 1-cycles, 2-cycles, and at most one 3-cycle.

Case 3. $|S_1| + |T_1| \leqslant n-1$.

Then A does not have any zero submatrix with total size greater than n, so $MR(A) = n$. By avoiding using possible non-zero entries in a suitable submatrix of $A[S_1^c, T_1^c]$ of total size less than n if necessary, we may assume that $m \geqslant 2$. Since the sum of the total sizes of all the maximal zero submatrices of A is at most $2n$ and $n \geqslant 4$, there are at most two maximal zero submatrices of A with total size $n-1$.

Subcase 3.1. $S_1 = T_1$. Note that $|S_k| + |T_k| \leqslant n-1$, for each $k = 1, 2, \cdots, m$. Take $i \in S_1$ and $j \in T_2 \subseteq T_1^c = S_1^c$. Since a_{ij} is an element of the full matrix $A[S_1, T_1^c]$ and a_{ji} is an element of the full matrix $A[S_1^c, T_1]$, we see that $a_{ij}a_{ji}$ is a 2-cycle of A. Note that A has at most two maximal zero submatrices of total size $n-1$. Upon deleting i th and j th rows and columns of A, each of the two maximal zero submatrices $A[S_1, T_1]$ and $A[S_2, T_2]$ (with largest total sizes) loses at least one line, and the principal submatrix $A' = A[\{i, j\}^c]$ of order $n-2$ satisfies $MR(A') = n-2$, as every zero submatrix of A' has total size at most $n-2$. By the induction hypothesis, A' has a composite cycle γ of length $c(A') = n-2$ consisting of 1-cycles, 2-cycles, and at most one 3-cycle. Thus $a_{ij}a_{ji}\gamma$ is a composite cycle of A of length $n = c(A)$ consisting of 1-cycles, 2-cycles, and at most one 3-cycle.

Subcase 3.2. $S_1 \neq T_1$. With the obvious modification that $m \geqslant 2$ instead of $m = 2$, the argument in Subcase 2.2 also works here.

Therefore, A has a composite cycle of length $n = c(A)$ consisting of 1-cycles, 2-cycles, and at most one 3-cycle.

The next result follows from Theorem 3.2.12 and Observation 3.2.7.

Theorem 3.2.13 Let A be an $n \times n$ sign pattern such that $mr(A) = 2$ and A has no zero line. Then A has a composite cycle of length $c(A)$ consisting of disjoint simple cycles of lengths up to 3, at most one of which is a 3-cycle.

Obviously, in the two preceding theorems, if $c(A)$ is odd and A has no 1-cycle, then A has a composite cycle of length $c(A)$ consisting of 2-cycles and exactly one 3-cycle.

In view of Theorem 3.2.13 and Theorem 3.1.14, we obtain the following result on the ranks achieved by diagonalizable matrices in the qualitative class of sign pattern matrix A such that $mr(A) = 2$ and A has no zero line.

Theorem 3.2.14 Let A be a square sign pattern.

(a) Suppose that $\gamma_1\gamma_2\cdots\gamma_k (k \geqslant 2)$ is a composite cycle of A of length $c(A)$ such that γ_1 is a composite cycle of length 2 that supports a rank-principal certificate for A, γ_2 is a 1-cycle, and for each $2 \leqslant i \leqslant k$, γ_1 is a 1-cycle or 2-cycle. Then $\{r \mid A \text{ allows diagonalizability with rank } r\} = \{2, 3, \cdots, c(A)\}$.

(b) More generally, suppose that $\gamma_1\gamma_2\cdots\gamma_k (k \geqslant 2)$ is a composite cycle of A where γ_1 is a composite cycle of length l_1 that supports a rank-principal certificate for A, and $\gamma_2, \gamma_3, \cdots, \gamma_k$ are simple cycles. Then

$\{l_1 + \sum_{j \in S} length(\gamma_i) \mid S \subseteq \{2, 3, \cdots, k\}\} \subseteq \{r \mid A$ allows diagonalizability with rank $r\}$.

Example 3.2.15 The sign pattern $A = \begin{bmatrix} 0 & - & - & - & - \\ - & 0 & - & - & - \\ + & + & 0 & + & + \\ + & + & 0 & + & + \\ + & + & 0 & + & + \end{bmatrix}$ satisfies $mr(A) = 2$ and A has no zero line. Note that the 2-cycle $a_{23}a_{32}$ supports a rank-principal certificate for $\boldsymbol{A}$ and $(a_{23}a_{32})(a_{15}a_{51})\ a_{44}$ is a composite cycle of length $c(\boldsymbol{A})=5$. By the preceding theorem, the set of the ranks of the diagonalizable matrices in $Q(\boldsymbol{A})$ is equal to $\{2, 3, 4, 5\}$.

Example 3.2.16 The sign pattern $\boldsymbol{A}=\begin{bmatrix} 0 & 0 & 0 & - & - & - \\ 0 & 0 & 0 & - & - & - \\ 0 & 0 & 0 & - & - & - \\ + & + & + & 0 & 0 & 0 \\ + & + & + & 0 & 0 & 0 \\ + & + & + & 0 & 0 & 0 \end{bmatrix}$ satisfies $mr(\boldsymbol{A})=2$ and $\boldsymbol{A}$ has no zero line. Note that the 2-cycle $a_{34}a_{43}$ supports a rank-principal certificate for $\boldsymbol{A}$ and $(a_{34}a_{43})\ (a_{25}a_{52})\ (a_{16}a_{61})$ is a composite cycle of length $c(\boldsymbol{A})=6$. By the preceding theorem, the set of the ranks of the diagonalizable matrices in $Q(\boldsymbol{A})$ contains $\{2, 4, 6\}$. But since the digraph of $\boldsymbol{A}$ is bipartite, every composite cycle of $\boldsymbol{A}$ has even length. Thus the set of the ranks of the diagonalizable matrices in $Q(\boldsymbol{A})$ is equal to $\{2, 4, 6\}$.

Example 3.2.17 The sign pattern $\boldsymbol{A}=\begin{bmatrix} 0 & - & - & - & - \\ + & 0 & - & - & - \\ + & + & 0 & - & - \\ + & + & + & 0 & - \\ + & + & + & + & 0 \end{bmatrix}$ satisfies $mr(\boldsymbol{A})=2$ since the polynomial sign change number (see [30]) of each row is 1. Note that the 2-cycle $a_{12}a_{21}$ supports a rank-principal certificate for $\boldsymbol{A}$ and $(a_{12}a_{21})(a_{34}a_{43})$ is a composite cycle of $\boldsymbol{A}$. By the preceding theorem, the set of the ranks of the diagonalizable matrices in $Q(\boldsymbol{A})$ contains $\{2, 4\}$. Also, the 3-cycle $a_{12}a_{23}a_{31}$ supports a rank-principal certificate for $\boldsymbol{A}$ and $(a_{12}a_{23}a_{31})\ (a_{45}a_{54})$ is a composite cycle of $\boldsymbol{A}$. By the preceding theorem, the set of the ranks of the diagonalizable matrices in $Q(\boldsymbol{A})$ contains $\{3, 5\}$. Thus the set of the ranks of the diagonalizable matrices in $Q(\boldsymbol{A})$ is equal to $\{2, 3, 4, 5\}$.

We conclude the paper with the following interesting problem.

Problem 3.2.18 Suppose that a sign pattern $\boldsymbol{A}$ allows diagonalizability and $mr(\boldsymbol{A})=2$. Let $\{r_1, r_2, \cdots, r_t\}=\{r \mid \boldsymbol{A}$ allows diagonalizability with rank $r\}$, where $r_1<r_2<\cdots<r_t$. Does it follow that $r_{j+1}-r_j\leqslant 2$ for each $j=1, 2, \cdots, t-1$?

Chapter 4

The Sign K-potent Sign Pattern Matrix That Allows Diagonalizability

Lee and Jin[28] considered sign idempotent sign pattern matrices which allow idempotent, so we will further study sign k-potent sign pattern matrix that allows sign k-potent in this chapter. Moreover, sign pattern allows k-potent can lead to diagonalizability. Finally, we further considered the relative between that sign k-potent sign pattern allows k-potent and that sign k-potent sign pattern allows diagonalizability.

The chapter is organized as follows. In Section 4. 1, two necessary and sufficient conditions allowing diagonalizability of a sign pattern are obtained. In Section 4. 2, we consider sign idempotent sign pattern matrix which allows diagonalizability. In Section 4. 3, that sign k-potent sign pattern matrix allows diagonalizability is considered.

4.1 Necessary and sufficient conditions of allowing diagonalizability

From Lemma 2. 2. 2. 3, we can easily get a necessary condition of allowing diagonalizability. The result is listed as the following:

Lemma 4. 1. 1 A sign pattern $\boldsymbol{A} \in Q_n$ allows diagonalizability, then at least there exists a composite cycle of length k in $\boldsymbol{A}$ with $mr(\boldsymbol{A}) \leqslant k \leqslant MR(\boldsymbol{A})$.

Proof In term of Lemma 2. 2. 2. 3, allowing diagonalizability iff there exists a real matrix $\boldsymbol{B} \in Q(\boldsymbol{A})$ with $rank(\boldsymbol{B}) = k$, such that $\boldsymbol{B}$ has a non-singular $k \times k$ principal submatrix, which can be denoted $\boldsymbol{C}$. Because $\boldsymbol{C}$ is a inverse matrix, the determinant of $\boldsymbol{C}$ is not equal to 0. Then at least one of terms of expansion of the determinant of $\boldsymbol{C}$ is not equal to 0. This term can be consist of a composite cycle of length k in $\boldsymbol{A}$. Therefore, the lemma holds.

Lemma 4. 1. 2 A composite cycle of length k in $\boldsymbol{A}$ can consist of a $k \times k$ principal submatrix. Converse to be also true.

Proof Suppose γ is composite cycle of length k in $\boldsymbol{A}$, denoted as $a_{i_1j_1}a_{i_2j_2}\cdots a_{i_kj_k}$, $i_1 \neq i_2 \neq \cdots \neq i_k \in \{1, 2, \cdots, n\}$, $j_1 \neq j_2 \neq \cdots \neq j_k \in \{1, 2, \cdots, n\}$, and $i_1, i_2, \cdots, i_k$ is equal to $j_1, j_2, \cdots, j_k$ one by one. Now we take a matrix $\boldsymbol{B} \in Q(\boldsymbol{A})$, such that each of $b_{i_1j_1}$, $b_{i_2j_2}$, $\cdots$, $b_{i_kj_k}$ is very large, the other entries in $\boldsymbol{B}$ are very small. The submatrix is denoted as $\boldsymbol{B}_0$, which is consisted of entries of $\boldsymbol{B}$ whose rows and columns both are i_1, i_2, $\cdots$, i_k. In terms of definition of determinant, we know $b_{i_1j_1}b_{i_2j_2}\cdots b_{i_kj_k}$ is a term of expansion of the determinant of $\boldsymbol{B}_0$, and the sum of the others terms is very small than $b_{i_1j_1}b_{i_2j_2}\cdots b_{i_kj_k}$. So $\boldsymbol{B}_0$ is a inverse submatrix.

Converse to be also true, which is obvious.

In term of Lemma 2.2.2.3 and Lemma 4.1.1, we can get the following theorem.

Theorem 4.1.3 If a sign pattern $\boldsymbol{A} \in Q_n$ and $mr(\boldsymbol{A}) = MR(\boldsymbol{A}) = k$, then $\boldsymbol{A}$ allows diagonalizability if and only if there exists some composite cycle of length k in $\boldsymbol{A}$.

Proof Necessity is obvious from Lemma 4.1.1.

Sufficiency There exists composite cycle of length k in A that can make us find a real $\boldsymbol{B} \in Q(\boldsymbol{A})$, which has a nonsingular $k \times k$ principal submatrix, denoted as $\boldsymbol{C}$. Due to $mr(\boldsymbol{A}) = MR(\boldsymbol{A}) = \mathrm{k}$, $rank(\boldsymbol{B}) = k$. According to Lemma 2.2.2.3, A allows diagonalizability. Therefore, the theorem holds.

According to Lemma 2.1.7, we know that changing a matrix by permutational similarity do not change diagonalizability. So now we modify the sign pattern $\boldsymbol{A}$. For a sign pattern $\boldsymbol{A}$, we can use permutation to made matrix $\boldsymbol{A}$ change as the following

$$\begin{bmatrix} \boldsymbol{A}_{11} & \boldsymbol{A}_{12} & 0 \\ \boldsymbol{A}_{21} & \boldsymbol{A}_{22} & \boldsymbol{A}_{23} \\ 0 & \boldsymbol{A}_{32} & \boldsymbol{A}_{33} \end{bmatrix},$$

and denote

$$\boldsymbol{B} = \boldsymbol{A}_{11},\ \boldsymbol{C} = \begin{bmatrix} \boldsymbol{A}_{11} & \boldsymbol{A}_{12} \\ \boldsymbol{A}_{21} & \boldsymbol{A}_{22} \end{bmatrix},$$

where $\boldsymbol{A}_{11}$ is $k \times k$, $\boldsymbol{A}_{22}$ is a square matrix, and $\boldsymbol{A}_{11}$ include a composite cycle of length k (Of course, if $\boldsymbol{A}$ have composite cycle, we will divide $\boldsymbol{A}$. If no, $\boldsymbol{A}$ does not allow diagonalizability. So we need not divide $\boldsymbol{A}$ again. Moreover, in some case $\boldsymbol{A}_{22}$, $\boldsymbol{A}_{32}$, $\boldsymbol{A}_{23}$ and $\boldsymbol{A}_{33}$ may be not exist, or might also have several kind of partition forms). For convenience, without loss of generality, we make the sign pattern be modified as the above case in the following theorem.

Theorem 4.1.4 A sign pattern $\boldsymbol{A} \in Q_n$, $\boldsymbol{A}$ allows diagonalizability if and only if there exists a kind of partition of $\boldsymbol{A}$, such that $\boldsymbol{A}_{33} = 0$, $\boldsymbol{A}_{23} = 0$, and $\boldsymbol{A}_{32} =$

0, moreover, $\exists$ real matrix $\boldsymbol{B}_0 \in Q(\boldsymbol{B})$, $\boldsymbol{C}_0 \in Q(\boldsymbol{C})$, such that $rank(\boldsymbol{B}_0) = rank(\boldsymbol{C}_0) = k$.

Proof Necessity. If $\boldsymbol{A}$ allows diagonalizability, according to Lemma 2. 2. 2. 3, there exists a real matrix $\boldsymbol{A}_0 \in Q(\boldsymbol{A})$ with $rank(\boldsymbol{A}_0) = k$, such that $\boldsymbol{A}_0$ has a non-singular $k \times k$ principal submatrix. By Lemma 4. 1. 2, the non-singular $k \times k$ principal submatrix can make a composite cycle of length k in $\boldsymbol{A}$. Without loss of generality, the entries which are made by composite cycle of length k in $\boldsymbol{A}$ can be written as $\boldsymbol{A}_{11}$. In term of Lemma 2. 2. 2. 3, if $\boldsymbol{A}_{33} \neq 0$ or $\boldsymbol{A}_{23} \neq 0$ or $\boldsymbol{A}_{32} \neq 0$ or $\nexists$ real matrix $\boldsymbol{B}_0 \in Q(\boldsymbol{B})$, $\boldsymbol{C}_0 \in Q(\boldsymbol{C})$ such that $rank(\boldsymbol{B}_0) = rank(\boldsymbol{C}_0) = k$, then we could not find any real matrix which can diagonalizability.

Sufficiency If $\boldsymbol{A}_{33} = 0$, $\boldsymbol{A}_{23} = 0$, $\boldsymbol{A}_{32} = 0$, and $\exists$ real matrix $\boldsymbol{B}_0 \in Q(\boldsymbol{B})$, $\boldsymbol{C}_0 \in Q(\boldsymbol{C})$ such that $rank(\boldsymbol{B}_0) = rank(\boldsymbol{C}_0) = k$, according to Lemma 2. 2. 2. 3, we can find out a non-singular $k \times k$ principal submatrix of some real matrix whose sign pattern is $\boldsymbol{A}$. So $\boldsymbol{A}$ allows diagonalizability.

4. 2 Sign idempotent sign pattern matrix

Lee and Jin[28] obtained some results that sign idempotent sign pattern matrices can allow idempotent. In our book, we mainly are concerned about diagonalizability. Meanwhile, we can also easily discover the relative of sign pattern matrices between allowing idempotent and allowing diagonalizability.

Theorem 4. 2. 1 If a sign pattern $\boldsymbol{A} \in Q_n$, $\boldsymbol{A}$ allows idempotent, then it allows diagonalizability.

The reducible sign idempotent sign pattern class in which $\boldsymbol{A}_{ij} = 0$ whenever $\boldsymbol{A}_{ii}$ and $\boldsymbol{A}_{jj}$ are entrywise positive is denoted by PPO.

Lemma 4. 2. 2[29] Let $\boldsymbol{A}$ be a reducible sign idempotent sign pattern matrix in Frobenius normal form, then $\boldsymbol{A}$ allows idempotent if and only if $\boldsymbol{A}$ is in PPO.

So we can easily get the following result.

Theorem 4. 2. 3 Let $\boldsymbol{A}$ be a reducible sign idempotent sign pattern matrix in Frobenius normal form, and $\boldsymbol{A}$ is also in PPO, then $\boldsymbol{A}$ allows diagonalizability.

4.3 Sign pattern matrices that are power

By using the Jordan canonical form, it is easy to see that for any positive integer m, a square non-singular real matrix $\boldsymbol{B}$ is diagonalizable if and only if $\boldsymbol{B}^m$ is diagonalizable. For sign pattern matrices, the result is not hold. If $\boldsymbol{A}$ allows diagonalizability, then $\boldsymbol{A}^m$ allows diagonalizability; But $\boldsymbol{A}^m$ allows diagonalizability, $\boldsymbol{A}$ might not be allow diagonalizability.

Based on this, in this section, we want to find out the relative of the above referred. Meanwhile we also only consider the case sign pattern matrices are power.

Lemma 4.3.1[39] Let $\boldsymbol{A}$ be an irreducible, sign k-potent pattern matrix with block $m \times m$ cyclic $\hat{\boldsymbol{A}}$, form given with $m \geqslant 1$. For $1 \leqslant i \leqslant m$, let $\boldsymbol{J}_i$ denote the matrix of pluses that has the same size as $\boldsymbol{A}_i$; and let $\boldsymbol{C}_i$ denote the first column of $\boldsymbol{A}_i$. Then: $m \in \{k/2, k\}$, and there exist vectors $\boldsymbol{u}_i$ in $\{+\boldsymbol{C}_i, -\boldsymbol{C}_i\}$ for $1 \leqslant i \leqslant m$ such that $\boldsymbol{A}_i = \boldsymbol{u}_i \boldsymbol{u}_{i+1}^{\mathrm{T}}$ for $1 \leqslant i \leqslant m-1$ (when $m \geqslant 2$), and $\boldsymbol{A}_m = \alpha \boldsymbol{u}_m \boldsymbol{u}_1^{\mathrm{T}}$ where $\alpha = +$ when $m = k$, and $\alpha = -$ when $m = k/2$. Further, if $\boldsymbol{D} = diag(\boldsymbol{u}_1, \boldsymbol{u}_2, \cdots, \boldsymbol{u}_m)$, then

$$\boldsymbol{D}\hat{\boldsymbol{A}}\boldsymbol{D}^{-1} = \begin{bmatrix} 0 & \boldsymbol{J}_1 & 0 & 0 & \cdots & 0 \\ 0 & 0 & \boldsymbol{J}_2 & 0 & \cdots & 0 \\ 0 & 0 & 0 & \boldsymbol{J}_3 & \cdots & 0 \\ \vdots & \vdots & \vdots & \vdots & \ddots & \vdots \\ 0 & 0 & 0 & 0 & \cdots & \boldsymbol{J}_{m-1} \\ \alpha \boldsymbol{J}_m & 0 & 0 & 0 & \cdots & 0 \end{bmatrix}.$$

By term of Lemma 4.3.1, we can get the following results.

Theorem 4.3.2 If $\boldsymbol{A}$ is an irreducible sign k-potent sign pattern matrix, then $\boldsymbol{A}$ allows sign k-potent.

Proof If $\boldsymbol{A}$ is an irreducible sign k-potent sign pattern matrix, and suppose that $\boldsymbol{A}$ has index of imprimitivity m for some $m \geqslant 2$, there exists permatation $\boldsymbol{P}$, such that $\boldsymbol{A}$ can be written as Frobenius normal form by permutation similarity, denoted as $\boldsymbol{P}^{\mathrm{T}}\boldsymbol{A}\boldsymbol{P} = \hat{\boldsymbol{A}}$. By Lemma 4.3.1, $\exists \boldsymbol{D}$, such that $\hat{\boldsymbol{A}}$ is similar to

$$\begin{bmatrix} 0 & \boldsymbol{J}_1 & 0 & 0 & \cdots & 0 \\ 0 & 0 & \boldsymbol{J}_2 & 0 & \cdots & 0 \\ 0 & 0 & 0 & \boldsymbol{J}_3 & \cdots & 0 \\ \vdots & \vdots & \vdots & \vdots & \vdots & \vdots \\ 0 & 0 & 0 & 0 & \cdots & \boldsymbol{J}_{m-1} \\ \alpha \boldsymbol{J}_m & 0 & 0 & 0 & \cdots & 0 \end{bmatrix}.$$

Let

$$\boldsymbol{B}_{i_0 j_0} = \begin{cases} \dfrac{1}{n_i}, & n_1 + \cdots + n_{i-1} \leqslant i_0 \leqslant n_1 + \cdots + n_i,\ n_2 + \cdots + n_i \leqslant j_0 \leqslant n_2 + \cdots + n_{i+1}. \\ & \text{and when } i = m,\ n_{i+1} = n_1. \\ 0, & \text{other.} \end{cases}$$

where $\boldsymbol{J}_i$, $i = 1, 2, \cdots, m$ is $n_i \times n_{i+1}$, J_m is $n_m \times n_1$, then $\boldsymbol{B}$ is sign k-potent, and $\exists \boldsymbol{P}_0 \in Q(\boldsymbol{P})$, $\boldsymbol{D}_0 \in Q(\boldsymbol{D})$, such that $\boldsymbol{P}_0^{\mathrm{T}} \boldsymbol{D}_0^{-1} \boldsymbol{B} \boldsymbol{D}_0 \boldsymbol{P}_0 \in Q(\boldsymbol{A})$ is also sign k-potent. So $\boldsymbol{A}$ allows sign k-potent.

Similar to Theorem 4.3.2, $\boldsymbol{A}$ allows sign k-potent, obviously $\boldsymbol{A}$ also allows diagonalizability. So we can get:

Colollary 4.3.3 If $\boldsymbol{A}$ is an irreducible sign k-potent sign pattern matrix, then $\boldsymbol{A}$ allows diagonalizability.

In the following section we will consider reducible sign k-potent sign pattern matrix, and obtain the following result.

Theorem 4.3.4 Let $\boldsymbol{A}$ be a reducible sign k-potent sign pattern matrix in Frobenius normal form with no zero diagonal entries. Then $\boldsymbol{A}$ allows sign k-potent if and only if $\boldsymbol{A}$ is the direct sum of irreducible sign k-potent sign pattern matrices.

Proof If $\boldsymbol{A}$ is an reducible sign pattern matrix in modified Frobenius normal form with no zero diagonal entries, then

$$\boldsymbol{A}^{k+1} = \begin{bmatrix} A_{11} & A_{12} & \cdots & A_{1n} \\ 0 & A_{22} & \cdots & A \\ \vdots & \vdots & \vdots & \vdots \\ 0 & 0 & \cdots & A_{mm} \end{bmatrix}^{k+1} = \begin{bmatrix} A_{11} & A_{12} & \cdots & A_{1n} \\ 0 & A_{22} & \cdots & A \\ \vdots & \vdots & \vdots & \vdots \\ 0 & 0 & \cdots & A_{mm} \end{bmatrix}.$$

And in terms of matrix multiplication, we have

$$A_{ij} = A_{ii}{}^k A_{ij} + A_{ii}{}^{k-1} A_{ij} A_{jj} + \cdots,\ i<j,\ i, j \in \{1, 2, \cdots, n\}.$$

Left multiply A_{ii} in two sides of above equation, we can get

$$A_{ii}A_{ij} = A_{ii}^{k+1}A_{ij} + A_{ii}^{k}A_{ij}A_{jj} + \cdots$$

If A allows sign k-potent, then there exists real matrix $B \in Q(\boldsymbol{A})$ such that

$$B_{ii}B_{ij} = B_{ii}^{k+1}B_{ij} + B_{ii}^{k}B_{ij}B_{jj} + \cdots.$$

B is a reducible k-potent matrix, then we also have $B_{ii}^{k+1} = B_{ii}$. Transposition of terms,

$$B_{ii}^{k}B_{ij}B_{jj} + \cdots = 0.$$

$A_{ii}A_{ij}$ is not umambiguous, so every terms of left side of above equation should be 0, namely, $B_{ii}^{k}B_{ij}B_{jj} = 0$. So $A_{ii}^{k}A_{ij}A_{jj} = 0$. Moreover, because A_{ii} and A_{jj} are irreducible k-potent, we can get

$$A_{ii}^{k}A_{ij}A_{jj}^{k} = 0A_{jj}^{k-1} = IA_{ij}I = A_{ij} = 0.$$

So the result holds.

Theorem 4.3.5 Let A be a reducible sign k-potent sign pattern matrix in Frobenius normal form. Then A allows sign k-potent if and only if A is the direct sum of irreducible sign k-potent sign pattern matrices, or exist some zero blocks in main diagonal line, the others are irreducible sign k-potent sign pattern matrices.

Proof According to Theorem 4.3.4, we only need to prove the case exist some zero blocks in main diagonal line. Let

$$B_{i_0j_0} = \begin{cases} \dfrac{1}{n_i}, & n_1 + \cdots + n_{i-1} \leqslant i_0 \leqslant n_1 + \cdots + n_i,\ n_2 + \cdots + n_i \leqslant j_0 \leqslant n_2 + \cdots + n_{i+1},\ \text{when } i = m,\ n_{i+1} = n_1, \text{ and } A_{ii},\ A_{jj} \neq 0. \\ \dfrac{\sum\limits_{l=n_{i-1}+1}^{n_i} \operatorname{sgn}(a_{lj_0})}{n_i}, & n_1 + \cdots + n_{i-1} \leqslant i_0 \leqslant n_1 + \cdots + n_i,\ n_2 + \cdots + n_i \leqslant j_0 \leqslant n_2 + \cdots + n_{i+1},\ \text{when } i = m,\ n_{i+1} = n_1, \text{ and } A_{ii} \neq 0,\ A_{jj} = 0. \\ 0, & \text{others.} \end{cases}$$

where $\operatorname{sgn}(a) = \begin{cases} 1, & a > 0. \\ 0, & a = 0 \\ -1 & a < 0. \end{cases}$, then B is sign k-potent. So by term of Lemma. 4.3.1, $\exists D_0 \in Q(D)$, such that $D_0^{-1}BD_0 \in Q(A)$ is also sign k-potent. So the result holds.

Colollary 4.3.6 If A is an reducible sign k-potent sign pattern matrix and A

allows sign k-potent, then allows diagonalizability.

In this section, we only considered sign k-potent sign pattern matrix allows sign k-potent, then it allows diagonalizability. In the future, the relative of allowing diagonalizability between $\boldsymbol{A}$ and $\boldsymbol{A}^m$ still deserve to further study deeply.

Chapter 5

New Results on Sign Patterns That Allow Diagonalizability

5.1 Introduction and preliminaries

The search for sufficient and necessary conditions characterizing sign patterns that allow diagonalizability has been a long-standing open problem. In this chapter, we further investigate sign patterns that allow diagonalizability.

We now introduce some more definitions and notation, most of which can be found in [14, 21, 37].

The largest possible length of the composite cycles of $\boldsymbol{A}$ is called the maximum cycle length of $\boldsymbol{A}$, denoted by $c(\boldsymbol{A})$. If $\boldsymbol{A}$ has no simple cycle at all, then $c(\boldsymbol{A})=0$.

A formal product of non-zero entries of a not necessarily square sign pattern $\boldsymbol{A} = (a_{ij})$ of the form

$$a_{i_1j_1}a_{i_2j_2}\cdots a_{i_kj_k}$$

with distinct row indices $i_1, i_2, \cdots, i_k$ and distinct column indices $j_1, j_2, \cdots, j_k$ is called a matching of size k. We say that the matching M is a principal matching if $\{i_1, i_2, \cdots, i_k\} = \{j_1, j_2, \cdots, j_k\}$. We say that the matching M supports a submatrix $\boldsymbol{B}$ if the row index set and column index set of $\boldsymbol{B}$ are equal to those of M. A sub-matching of M is a matching consisting of some entries in M.

Observe that a composite cycle may be viewed as a principal matching, and vice versa.

It is clear that $MR(\boldsymbol{A})$ is equal to the maximum possible size of a matching of $\boldsymbol{A}$. But in fact it is very difficult to determine $mr(\boldsymbol{A})$. An $n\times n$ sign pattern whose minimum rank equals n is said to be sign non-singular.

Let $z(\boldsymbol{B})$ and $g(\boldsymbol{B})$ denote the algebraic and geometric multiplicities of 0 as an eigenvalue of real matrix $\boldsymbol{B}$. Moreover, for a square sign pattern matrix $\boldsymbol{A}$, $z(\boldsymbol{A}) = \min\{z(\boldsymbol{B}) \mid \boldsymbol{B}\in Q(\boldsymbol{A})\}$ denotes the minimum algebraic multiplicity of 0 as an eigenvalue of a matrix in $Q(\boldsymbol{A})$, $g(\boldsymbol{A})$ denotes the minimum geometric multiplicity of 0 as an eigenvalue of a matrix in $Q(\boldsymbol{A})$.

The chapter is organized as follows. In Section 4.2, some necessary and/or sufficient conditions for a sign pattern to allow diagonalizability are obtained. In Section 4.3, we present some results on how to change entries of a matrix to obtain another matrix with the same sign pattern and a prescribed rank. In Section 4.4, results on special types of sign pattern matrices that allow diagonalizability are considered. In Section 4.5, we further obtain some necessary and/or sufficient conditions for a sign pattern matrix in Frobenius normal form to allow diagonalizability, and partly solved out two problems in [10].

5.2 Necessary and/or sufficient conditions for allowing diagonalizability

Following [14], we use the following terminology.

Definition 5.2.1 A real matrix $\boldsymbol{B}$ is said to be rank-principal if $\boldsymbol{B}$ has a non-singular $k \times k$ principal submatrix $\boldsymbol{C}$, where $k = rank(\boldsymbol{B})$. Such a principal submatrix $\boldsymbol{C}$ is called a rank-principal certificate of $\boldsymbol{B}$.

Definition 5.2.2 We say that a composite cycle γ of a square sign pattern $\boldsymbol{A}$ supports a rank-principal certificate for $\boldsymbol{A}$ if there exists a real matrix $\boldsymbol{B} \in Q(\boldsymbol{A})$ that is rank-principal and the index set of γ is equal to the row index set of a rank-principal certificate of $\boldsymbol{B}$.

Remark 5.2.3 As pointed out in [14], the only possible value of k in Lemma 2.2.2.8 is $k = mr(\boldsymbol{A})$, and every chordless composite cycle of length $k = mr(\boldsymbol{A})$ of $\boldsymbol{A}$ supports a rank-principal certificate for $\boldsymbol{A}$, which ensures that $\boldsymbol{A}$ allows diagonalizability with rank $mr(\boldsymbol{A})$. However, some composite cycles with chords could support a rank-principal certificate, as the following example shows.

For

$$\boldsymbol{A} = \begin{bmatrix} 0 & + & + & 0 & + \\ 0 & 0 & + & 0 & + \\ 0 & 0 & 0 & + & + \\ + & 0 & 0 & 0 & + \\ + & + & + & + & + \end{bmatrix},$$

the cycle $\gamma = a_{12}a_{23}a_{34}a_{41}$ has a chord a_{13}, but γ supports a rank-principal certificate for $\boldsymbol{A}$. Thus $\boldsymbol{A}$ allows diagonalizability with rank 4.

For a sign pattern whose entries are from the set $\{0, +\}$ (or $\{0, -\}$, how many chords can we add to a cycle which still can support a rank principal certificate? In the above matrix, we can add 6 chords, and change it to the following matrix:

$$A = \begin{bmatrix} + & + & + & + & + \\ + & 0 & + & + & + \\ + & 0 & 0 & + & + \\ + & 0 & 0 & 0 & + \\ + & + & + & + & + \end{bmatrix}.$$

Moreover, it is easy to see the following results.

Theorem 5.2.4 Let $A \in Q_n$.

(1) (see [14]) If there exists a chordless composite cycle of length k that can support a rank-principal certificate for A, then A allows diagonalizability with rank k, and $k = mr(A)$;

(2) For a sign pattern whose entries are from the set $\{0, +\}$ (or $\{0, -\}$, if a composite cycle of length k can support a rank-principal certificate, then it has at most $\frac{k(k-1)}{2}$ chords.

Theorem 5.2.5 A sign pattern $A \in Q_n$ allows diagonalizability with rank k if and only if A allows a rank-principal matrix of rank k.

It is worth noting that set of sign patterns that allow diagonalizability is also closed under direct sums, and Kronecker products.

Theorem 5.2.6 Let $A \in Q_n$. Suppose that there exist two composite cycles γ_1 and γ_2 with $\gamma_1 \subset \gamma_2$. If γ_1 can support a rank-principal certificate, then γ_2 can also support a rank-principal certificate.

Theorem 5.2.7 A square sign pattern A allows diagonalizability with rank $MR(A)$ if and only if $c(A) = MR(A)$.

5.3 Entry modifications between matrices achieving extreme ranks

Ranks play an important role in many matrix problems such as diagonalizability. For a sign pattern A, we consider the minimum number of entries which need to be modified to obtain a matrix $B' \in Q(A)$ with $MR(A)$ starting with a matrix $B \in Q(A)$ with rank $mr(A)$. We observe the following result.

Theorem 5.3.1 Let $\boldsymbol{A}$ be a sign pattern. Starting with any matrix $\boldsymbol{B} \in Q(\boldsymbol{A})$ with $rank(\boldsymbol{B}) = mr(\boldsymbol{A})$, we may change at most $MR(\boldsymbol{A})$ entries of $\boldsymbol{B}$ to obtain a matrix $\boldsymbol{B}' \in Q(\boldsymbol{A})$ with rank $MR(\boldsymbol{A})$.

Proof It is known that $\boldsymbol{A}$ has a maximum matching M of size $MR(\boldsymbol{A})$. For any $\boldsymbol{B} \in Q(\boldsymbol{A})$ with $rank(\boldsymbol{B}) = mr(\boldsymbol{A})$, we can change the entries of $\boldsymbol{B}$ in M to very large values. Then the resulting matrix $\boldsymbol{B}' \in Q(\boldsymbol{A})$ has rank $MR(\boldsymbol{A})$.

In fact, it suffices to change $MR(\boldsymbol{A}) - 1$ entries of $\boldsymbol{B}$ in M in the preceding argument, as a dominant non-zero term in the determinant expansion of a submatrix of order $MR(\boldsymbol{A})$ may be created this way. This leads to the following result.

Corollary 5.3.2 Let $\boldsymbol{A}$ be a sign pattern. It suffices to change $MR(\boldsymbol{A}) - 1$ entries of any $\boldsymbol{B} \in Q(\boldsymbol{A})$ with $rank(\boldsymbol{B}) = mr(\boldsymbol{A})$ to get a matrix $\boldsymbol{B}' \in Q(\boldsymbol{A})$ with rank $MR(\boldsymbol{A})$.

The next result is on sign patterns whose minimum rank and maximum rank differ by 1.

Theorem 5.3.3 Let $\boldsymbol{A} = (a_{ij}) \in Q_n$ with $MR(\boldsymbol{A}) = mr(\boldsymbol{A}) + 1$. Then one can change just one entry of a suitable matrix $\boldsymbol{B} \in Q(\boldsymbol{A})$ with rank $mr(\boldsymbol{A})$ to get a matrix $\boldsymbol{B}' \in Q(\boldsymbol{A})$ with rank $MR(\boldsymbol{A})$.

Proof Let $m = MR(\boldsymbol{A})$ and let $M = a_{i_1 j_1} a_{i_2 j_2} \cdots a_{i_m j_m}$ be a matching of size m in $\boldsymbol{A}$. Let $\boldsymbol{B}_0 \in Q(\boldsymbol{A})$ be a matrix with $rank(\boldsymbol{B}_0) = mr(\boldsymbol{A}) = m - 1$. Replace the (i_1, j_1) entry of $\boldsymbol{B}_0$ by a much larger number of same sign so that the absolute value of the new entry is greater than the sum of the absolute values of the other entries in the same row of $\boldsymbol{B}_0$. Denote the resulting matrix by $\boldsymbol{B}_1$. We then replace the (i_2, j_2) entry of $\boldsymbol{B}_1$ by a much larger number of same sign to get a matrix $_2$. Proceeding this way, we get $\boldsymbol{B}_0, \boldsymbol{B}_1, \cdots, \boldsymbol{B}_m \in Q(\boldsymbol{A})$. Since $\boldsymbol{B}_m$ has m large entries that form a generalized dominant diagonal of a submatrix of order m, we see that $rank(\boldsymbol{B}_m) = m = MR(\boldsymbol{A})$. Let k be the smallest positive integer such that $rank(\boldsymbol{B}_m) = m$. Then we have rank $rank(\boldsymbol{B}_{k-1}) = m - 1$ and $rank(\boldsymbol{B}_m) = m$, and these matrices differ in one entry only. This completes the proof.

We now show a similar result when the gap between the minimum rank and the maximum rank may be larger.

Theorem 5.3.4 Let $\boldsymbol{A}$ be a sign pattern with $MR(\boldsymbol{A}) = m$ and $mr(\boldsymbol{A}) = r$. Suppose that there is a matrix $\boldsymbol{B} \in Q(\boldsymbol{A})$ with $rank(\boldsymbol{B}) = r$ such that an $r \times r$ non-singular submatrix $\boldsymbol{C}$ of $\boldsymbol{B}$ is supported by a submatching of a matching M of size m of $\boldsymbol{A}$. Then one can change only $m-r$ entries of $\boldsymbol{B}\in Q(\boldsymbol{A})$ to obtain a matrix $\boldsymbol{B}'\in Q(\boldsymbol{A})$ with $rank(\boldsymbol{B}')=MR(\boldsymbol{A})$.

Proof Permuting the rows and columns of $\boldsymbol{A}$ if necessary, without loss of generality, we may assume that $\boldsymbol{C}$ is in the upper left corner of $\boldsymbol{B}$, that is, $\boldsymbol{B}=\begin{bmatrix}\boldsymbol{C} & \boldsymbol{D}\\ \boldsymbol{E} & \boldsymbol{F}\end{bmatrix}$. The hypothesis ensures that $\boldsymbol{C}$ contains a matching M_1 of size r and $\boldsymbol{F}$ contains a matching M_2 of size $m-r$, where $M_1M_2=M$.

Since the rows (respectively, columns) of $\boldsymbol{B}$ containing $\boldsymbol{C}$ are linearly independent and can span the other rows (respectively, columns) of $\boldsymbol{B}$, we may use suitable elementary row and column operations to reduce $\boldsymbol{B}$ to $\begin{bmatrix}\boldsymbol{C} & 0\\ 0 & 0\end{bmatrix}$. If the entries of $\boldsymbol{F}$ in the matching M_2 are replaced with twice of the original values to get a matrix $\boldsymbol{F}'$, then when the above-mentioned elementary row and column operations are applied to the matrix $\boldsymbol{B}'=\begin{bmatrix}\boldsymbol{C} & \boldsymbol{D}\\ \boldsymbol{E} & \boldsymbol{F}'\end{bmatrix}\in Q(\boldsymbol{A})$, we get a matrix of the form $\begin{bmatrix}\boldsymbol{C} & 0\\ 0 & \boldsymbol{G}\end{bmatrix}$, where $\boldsymbol{G}$ has exactly $m-r$ non-zero entries, in the same positions as entries in M_2. It follows that $rank(\boldsymbol{B}')=m$, and $\boldsymbol{B}'$ is obtained from $\boldsymbol{B}$ by modifying exactly $m-r$ entries.

We suspect that the hypothesis of the preceding theorem is satisfied by every sign pattern matrix, which motivates the following conjecture.

Conjecture 5.3.5 For every sign pattern $\boldsymbol{A}$, it is always possible to change $MR(\boldsymbol{A})-mr(\boldsymbol{A})$ entries of some matrix $\boldsymbol{B}\in Q(\boldsymbol{A})$ with $rank(\boldsymbol{B})=mr(\boldsymbol{A})$ to get a matrix $\boldsymbol{B}'\in Q(\boldsymbol{A})$ with rank $MR(\boldsymbol{A})$.

We think the results of Theorem 5.3.4 in [10] is also necessity, so we have the following results.

Theorem 5.3.6 Let $\boldsymbol{A}$ be a sign pattern with $MR(\boldsymbol{A})=\mathrm{m}$ and $mr(\boldsymbol{A})=r$. If one can change only $m-r$ entries of $\boldsymbol{B}\in Q(\boldsymbol{A})$ to obtain a matrix $\boldsymbol{B}'\in Q(\boldsymbol{A})$ with $rank(\boldsymbol{B}')=MR(\boldsymbol{A})$, then there is an $r\times r$ non-singular submatrix $\boldsymbol{C}$ of $\boldsymbol{B}$ is supported by a submatching of a matching M of size m of $\boldsymbol{A}$.

Proof We know permutation of matrix do not change the rank of matrix. So permuting the rows and columns of $\boldsymbol{A}$ if necessary, without loss of generality, we may assume that $\boldsymbol{C}$ is in the upper left corner of $\boldsymbol{B}$, that is, $\boldsymbol{B}=\begin{bmatrix}\boldsymbol{C} & \boldsymbol{D}\\ \boldsymbol{E} & \boldsymbol{F}\end{bmatrix}$. The hypothesis ensures that $\boldsymbol{C}$ contains a matching M_1 of size r and $rank(\boldsymbol{C})=rank(\boldsymbol{B})=r$. Since the rows (respectively, columns) of $\boldsymbol{B}$ containing $\boldsymbol{C}$ are linearly independent and can span the other rows (respectively, columns) of $\boldsymbol{B}$, we may use suitable elementary row and column operations to reduce $\boldsymbol{B}$ to $\begin{bmatrix}\boldsymbol{C} & 0\\ 0 & 0\end{bmatrix}$. If we want the rank of $\boldsymbol{B}$ is bigger, we can change its entries of $\boldsymbol{B}$. The following we will consider it according to the place of entries of matrix $\boldsymbol{B}$.

(1) If we change someone entries f_{ij} as $f_{ij}+\varepsilon$ of F. Then through the same permutation of matrix to $\boldsymbol{B}$, we can get $\boldsymbol{B}$ as

$$\boldsymbol{B}'=\begin{bmatrix}\boldsymbol{C} & 0 & \cdots & 0 & \cdots & 0\\ 0 & 0 & \cdots & \cdots & \cdots & 0\\ \vdots & \vdots & \vdots & \vdots & \vdots & \vdots\\ 0 & 0 & \cdots & \varepsilon & \cdots & 0\\ \vdots & \vdots & \vdots & \vdots & \vdots & \vdots\\ 0 & 0 & \cdots & 0 & \cdots & 0\end{bmatrix}.$$

Obviously, $rank(\boldsymbol{B}')=r+1$.

(2) If we change someone entries d_{ij} as $d_{ij}+\varepsilon$ of $\boldsymbol{D}$ (or $\boldsymbol{E}$). Then similarly through some the same permutation of matrix to $\boldsymbol{B}$, we can get $\boldsymbol{B}$ as $\boldsymbol{B}'=\begin{bmatrix} & 0 & \cdots & 0 & \cdots & 0\\ & \vdots & \vdots & \vdots & \vdots & \vdots\\ C & 0 & \cdots & \varepsilon & \cdots & 0\\ & \vdots & \vdots & \vdots & \vdots & \vdots\\ & 0 & \cdots & 0 & \cdots & 0\\ E & & & 0 & & \end{bmatrix}$. If $rank(\boldsymbol{B}')=r+1$, then the rank of $\begin{bmatrix}\boldsymbol{C}\\ \boldsymbol{E}\end{bmatrix}$ delete the same row as d_{ij} in $\boldsymbol{B}$ should to be r. Therefore this case will transfer to case (1).

(3) If we change some entries c_{ij} as $c_{ij}+\varepsilon$ of $\boldsymbol{C}$. Then similarly through some the same permutation of matrix to $\boldsymbol{B}$, we can get $\boldsymbol{B}$ as

$$
B' = \begin{bmatrix}
c_{11} & \cdots & c_{1j} & \cdots & c_{1r} & 0 & \cdots & 0 & \cdots & 0 \\
\vdots & \vdots & \vdots & \vdots & \vdots & \vdots & \vdots & \vdots & \vdots & \vdots \\
c_{i1} & \cdots & c_{ij}+\varepsilon & \cdots & c_{ir} & 0 & \cdots & \varepsilon & \cdots & 0 \\
\vdots & \vdots & \vdots & \vdots & \vdots & \vdots & \vdots & \vdots & \vdots & \vdots \\
c_{r1} & \cdots & c_{rj} & \cdots & c_{rr} & 0 & \cdots & 0 & \cdots & 0 \\
 & & E & & & & & 0 & &
\end{bmatrix}.
$$

If $rank(\boldsymbol{B}') = r+1$, This case will transfer to case 2.

Now we can find that both case (2) and (3) transfer to case (1). Therefore, if we want to change only $m-r$ entries of $\boldsymbol{B} \in Q(\boldsymbol{A})$ to obtain a matrix $\boldsymbol{B}' \in Q(\boldsymbol{A})$ with $rank(\boldsymbol{B}') = MR(\boldsymbol{A})$, every time we must change the entry according to case (1). So the entries changed and some r entries of submatrix $\boldsymbol{C}$ of $\boldsymbol{B}$ will make a submatching of the matching M of size m of $\boldsymbol{A}$.

Moreover, we also have the similar results from the $MR(\boldsymbol{A})$ to $mr(\boldsymbol{A})$.

Corollary 5.3.7 Let $\boldsymbol{A}$ be a sign pattern with $MR(\boldsymbol{A}) = m$ and $mr(\boldsymbol{A}) = r$. We can change only $m-r$ entries of $\boldsymbol{B} \in Q(\boldsymbol{A})$ with $rank(\boldsymbol{B}) = MR(\boldsymbol{A})$ to obtain a matrix $\boldsymbol{B}' \in Q(\boldsymbol{A})$ with $rank(\boldsymbol{B}') = mr(\boldsymbol{A})$ if and only if there exists an $r \times r$ non-singular submatrix $\boldsymbol{C}$ of $\boldsymbol{B}'$ which is supported by a submatching of a matching M of size m of $\boldsymbol{A}$.

Obviously, the result also holds from any rank to another rank. Therefore, we have the following results:

Corollary 5.3.8 Let $\boldsymbol{A}$ be a sign pattern. We can change only $m-r$ entries of $\boldsymbol{B} \in Q(\boldsymbol{A})$ with $rank(\boldsymbol{B}) = m$ to obtain a matrix $\boldsymbol{B}' \in Q(\boldsymbol{A})$ with $rank(\boldsymbol{B}') = r$ if and only if there exists an $r \times r$ non-singular submatrix $\boldsymbol{C}$ of $\boldsymbol{B}'$ which is supported by a submatching of a matching M of size m of $\boldsymbol{A}$.

Therefore, according to Corollary 5.3.8, we can get the following results:

Theorem 5.3.9 Let $\boldsymbol{A}$ be a sign pattern. One can change only $m-r$ entries of $\boldsymbol{B} \in Q(\boldsymbol{A})$ with $rank(\boldsymbol{B}) = m$ to obtain a matrix $\boldsymbol{B}' \in Q(\boldsymbol{A})$ with $rank(\boldsymbol{B}') = r$, and there exists an $r \times r$ non-singular submatrix $\boldsymbol{C}$ of $\boldsymbol{B}'$ is supported by a submatching of a principal matching M of size $\leqslant m$ of $\boldsymbol{A}$, then $\boldsymbol{A}$ allows diagonalizability.

5.4 Further results on sign patterns that allow diagonalizability

For a square matrix $\boldsymbol{B}$, let $z(\boldsymbol{B})$ and $g(\boldsymbol{B})$ denote the algebraic and geometric multiplicities of 0 as an eigenvalue of $\boldsymbol{B}$. For a square sign pattern matrix A, $z(\boldsymbol{A}) = \min\{z(\boldsymbol{B}) \mid \boldsymbol{B} \in Q(\boldsymbol{A})\}$ denotes the minimum algebraic multiplicity of 0 as an eigenvalue of a matrix in $Q(\boldsymbol{A})$, and similarly, $Z(\boldsymbol{A})$ denotes the maximum algebraic multiplicity of 0 as an eigenvalue of a matrix in $Q(\boldsymbol{A})$, $g(\boldsymbol{A})$ denotes the minimum geometric multiplicity of 0 as an eigenvalue of a matrix in $Q(\boldsymbol{A})$, and $G(\boldsymbol{A})$ denotes the maximum geometric multiplicity of 0 as an eigenvalue of a matrix in $Q(\boldsymbol{A})$. The following result can be found in [40].

Theorem 5.4.1 Let $\boldsymbol{A}$ be an $n \times n$ sign pattern. If $z(\boldsymbol{A}) = g(\boldsymbol{A})$, then $\boldsymbol{A}$ allows diagonalizability.

Proof It can be seen that $z(\boldsymbol{A}) = n - c(\boldsymbol{A})$, and $g(\boldsymbol{A}) = n - MR(\boldsymbol{A})$. Hence, $z(\boldsymbol{A}) = g(\boldsymbol{A})$ ensures that $c(\boldsymbol{A}) = MR(\boldsymbol{A})$. Let $k = c(\boldsymbol{A})$.

Thus by emphasizing the entries on a composite cycle of length k, we get a matrix $\boldsymbol{B} \in Q(\boldsymbol{A})$ with $rank(\boldsymbol{B}) = k$ and $z(\boldsymbol{B}) = g(\boldsymbol{B}) = n - k$. Then the characteristics polynomial of $\boldsymbol{B}$ has the form

$$p_{\boldsymbol{B}}(t) = t^n - E_1 t^{n-1} + E_2 t^{n-2} - \cdots + (-1)^k E_k t^{n-k},$$

where $\boldsymbol{E}_i$, $i = 1, 2, \cdots, n-k$, is the sum of all i-by-i principal minors of $\boldsymbol{B}$, and $E_k \neq 0$. Thus, the matrix $\boldsymbol{B}$ is rank-principal. Therefore, $\boldsymbol{A}$ allows diagonalizability by Theorem 5.2.5.

As pointed out in the preceding proof, $z(\boldsymbol{A}) = g(\boldsymbol{A})$ is equivalent to $c(\boldsymbol{A}) = MR(\boldsymbol{A})$, which holds if and only if $\boldsymbol{A}$ allows diagonalizability with rank $MR(\boldsymbol{A})$ by Theorem 5.2.7.

We also have the following two results similar to Theorem 5.4.1.

Theorem 5.4.2 Let $\boldsymbol{A} \in Q_n$ be a sign pattern such that $Z(\boldsymbol{A}) = G(\boldsymbol{A})$. Then $\boldsymbol{A}$ allows diagonalizability with rank $mr(\boldsymbol{A})$.

Proof Take a matrix $\boldsymbol{B} \in Q(\boldsymbol{A})$ such that $rank(\boldsymbol{B}) = mr(\boldsymbol{A})$. Since $Z(\boldsymbol{A}) = G(\boldsymbol{A})$, we have

$$z(\boldsymbol{B}) \leqslant Z(\boldsymbol{A}) = G(\boldsymbol{A}) = n - mr(\boldsymbol{A}) = n - rank(\boldsymbol{B}).$$

Hence, $n - z(\boldsymbol{B}) \geqslant rank(\boldsymbol{B})$. But since the rank of any square matrix is clearly always greater than or equal to the number of non-zero eigenvalues of the matrix, we also have the opposite inequality $n - z(\boldsymbol{B}) \leqslant rank(\boldsymbol{B})$. Thus $n - z(\boldsymbol{B}) = rank(\boldsymbol{B})$, namely, the number of non-zero eigenvalues of $\boldsymbol{B}$ is equal to rank of $\boldsymbol{B}$. It follows that $\boldsymbol{B}$ is rank-principal. By Theorem 5.2.5, $\boldsymbol{A}$ allows diagonalizability with rank $mr(\boldsymbol{A})$.

As a square matrix for which the algebraic multiplicity and geometric multiplicity of the eigenvalue 0 are equal is rank-principal, by Theorem 5.2.5, we have the following fact.

Corollary 5.4.3 A square sign pattern $\boldsymbol{A}$ allows diagonalizability if and only if there exists a real matrix $\boldsymbol{B} \in Q(\boldsymbol{A})$ for which the algebraic multiplicity and geometric multiplicity of the eigenvalue 0 are equal.

Observe that up to permutational similarity, every square rank-principal matrix arises as a matrix of the form

$$\begin{bmatrix} \boldsymbol{C} & \boldsymbol{D} \\ \boldsymbol{E} & \boldsymbol{E}\boldsymbol{C}^{-1}\boldsymbol{D} \end{bmatrix},$$

where $\boldsymbol{C}$ is non-singular, and $\boldsymbol{D}$, $\boldsymbol{E}$ are of suitable sizes. Thus, up to permutational similarity, the sign patterns that allow diagonalizability are just sign patterns of matrices of this form. However, we are more interested in the combinatorial characterizations of sign patterns that allow diagonalizability.

We now present some special sign patterns that allow diagonalizability.

Theorem 5.4.4 Suppose a square sign pattern $\boldsymbol{A}$ has minimum rank $k>0$ and $\boldsymbol{A}$ has a sign non-singular $k \times k$ principal submatrix. Then $\boldsymbol{A}$ allows diagonalizability with rank k.

Proof Every matrix $\boldsymbol{B} \in Q(\boldsymbol{A})$ with *rank* k is clearly rank-principal due to the presence of a sign non-singular $k \times k$ principal submatrix of $\boldsymbol{A}$. Thus $\boldsymbol{A}$ allows diagonalizability with *rank* k by Theorem 5.2.5.

Next, we give a characterization of the square sign patterns that require a unique rank and allow diagonalizability.

Theorem 5.4.5 Let $\boldsymbol{A}$ be a square sign pattern such that $mr(\boldsymbol{A}) = MR(\boldsymbol{A}) =$ k. Then $\boldsymbol{A}$ allows diagonalizability if and only if $c(\boldsymbol{A}) = k$.

Proof Both the necessity and the sufficiency follow from Theorem 5.2.7.

Upper triangular sign patterns that allow diagonalizability are identified below.

Theorem 5.4.6 Let $\boldsymbol{A}$ be an upper triangular square sign pattern. Then $\boldsymbol{A}$ allows diagonalizability if and only if $c(\boldsymbol{A}) = mr(\boldsymbol{A})$.

Proof Since $\boldsymbol{A}$ is an upper triangular square sign pattern, every matrix $\boldsymbol{B} \in Q(\boldsymbol{A})$ has precisely $c(\boldsymbol{A})$ non-zero eigenvalues, so $mr(\boldsymbol{A}) \geqslant c(\boldsymbol{A})$.

Suppose that $\boldsymbol{A}$ allows diagonalizability, then $c(\boldsymbol{A}) \geqslant mr(\boldsymbol{A})$. In view of the opposite inequality above, we get $c(\boldsymbol{A}) = mr(\boldsymbol{A})$.

Conversely, assume that $c(\boldsymbol{A}) = mr(\boldsymbol{A})$. Let $\boldsymbol{B} \in Q(\boldsymbol{A})$ be such that

$rank(\boldsymbol{B}) = mr(\boldsymbol{A})$. Clearly, there is a diagonal matrix $\boldsymbol{D}$ with positive diagonal entries such that all the non-zero diagonal entries of $\boldsymbol{DB} \in Q(\boldsymbol{A})$ are distinct. Thus every nonzero eigenvalue of $\boldsymbol{DB}$ has algebraic and geometric multiplicity 1. If 0 is an eigenvalue of $\boldsymbol{DB}$, then its algebraic and geometric multiplicities are both equal to $n - c(\boldsymbol{A}) = n - mr(\boldsymbol{A}) = n - rank(\boldsymbol{DB})$. Hence, $\boldsymbol{DB} \in Q(\boldsymbol{A})$ is diagonalizable, so that $\boldsymbol{A}$ allows diagonalizability.

A square sign pattern is said to be idempotent if $\boldsymbol{A}^2$ is unambiguously defined, and $\boldsymbol{A}^2 = \boldsymbol{A}$. More generally, we say a sign pattern is k-potent (where k is a positive integer) if $\boldsymbol{A}^{1+k}$ is unambiguously defined and $\boldsymbol{A}^{1+k} = \boldsymbol{A}$. Such sign patterns always allow diagonalizability.

Theorem 5.4.7 Every sign k-potent sign pattern $\boldsymbol{A}$ allows diagonalizability with rank $mr(\boldsymbol{A})$.

Proof Let $\boldsymbol{A}$ be a k-potent sign pattern and let $B \in Q(\boldsymbol{A})$ be such that $rank(\boldsymbol{B}) = mr(\boldsymbol{A})$. On the one hand, clearly $rank(\boldsymbol{B}^{1+k}) \leqslant rank(\boldsymbol{B})$. On the other hand, since $rank(\boldsymbol{B}) = mr(\boldsymbol{A})$ and $\boldsymbol{B}^{1+k} \in Q(\boldsymbol{A}^{1+k}) = Q(\boldsymbol{A})$, we also have $rank(\boldsymbol{B}^{1+k}) \geqslant rank(\boldsymbol{B})$. Thus, $rank(\boldsymbol{B}^{1+k}) = rank(\boldsymbol{B})$. It follows that $rank(\boldsymbol{B}) = rank(\boldsymbol{B}^2) = \cdots = rank(\boldsymbol{B}^{1+k})$. By considering the Jordan canonical form of $\boldsymbol{B}$, we see that either $\boldsymbol{B}$ is non-singular or the eigenvalue 0 of $\boldsymbol{B}$ has index 1. Thus $rank(\boldsymbol{B})$ is equal to the number of non-zero eigenvalues of $\boldsymbol{B}$, which ensures that $\boldsymbol{B}$ is rank-principal. By Theorem 5.2.5, $\boldsymbol{A}$ allows diagonalizability with rank $mr(\boldsymbol{A})$.

Now we can get another sufficient condition of allowing diagonalizability.

Theorem 5.4.8 Let $\boldsymbol{A} \in Q_n$. Suppose that there exist a real matrix $\boldsymbol{B} \in Q(\boldsymbol{A})$ whose rank is $mr(\boldsymbol{A})$, and $\boldsymbol{B}$ has a non-zero minor N with order $mr(\boldsymbol{A})$, if there exist $mr(\boldsymbol{A})$ entries in N which is not at the same row and column (make one matching), and these entries add the others entries out of $\boldsymbol{N}$ in $\boldsymbol{B}$ can make one composite cycle (principal matching), then $\boldsymbol{A}$ allows diagonalizability.

Proof Generally speaking, in order to expression, we can choose $\boldsymbol{B} = \begin{bmatrix} \boldsymbol{N} & * \\ * & \boldsymbol{M} \end{bmatrix}$. Denote the composite cycle as γ, which are made by $mr(\boldsymbol{A})$ entries in $\boldsymbol{N}$ and $\boldsymbol{P}$ in $\boldsymbol{M}$. Let

$$M=\begin{bmatrix} m_{11} & m_{12} & \cdots & m_{1t} \\ m_{21} & m_{22} & \cdots & m_{2t} \\ \vdots & \vdots & \vdots & \vdots \\ m_{t1} & m_{t1} & \cdots & m_{tt} \end{bmatrix}.$$

If m_{12} is in γ, then we can change m_{12} as $m_{12}+\varepsilon$. Thus, we can use elementary transformation make $\boldsymbol{B}$ as $\boldsymbol{B}'=\begin{bmatrix} \boldsymbol{N} & 0 & \cdots & 0 \\ 0 & \varepsilon & \cdots & 0 \\ \vdots & \vdots & \vdots & \vdots \\ 0 & 0 & \cdots & 0 \end{bmatrix}$. Therefore, the rank of $\boldsymbol{B}'$ is $mr(\boldsymbol{A})+1$.

Similarly, we can change the other $p-1$ entries which is in $\boldsymbol{M}$ and also in γ. Finally, we can find the rank of new matrix is $mr(\boldsymbol{A})+p$ and the composite γ can make a rank-principal certificate. Therefore, the new matrix is rank-principal.

For convenience, we present the following definition:

Definition 5.4.9 Let $\boldsymbol{A}\in Q_n$. Suppose that there exist a real matrix $\boldsymbol{B}\in Q(\boldsymbol{A})$ whose rank is $mr(\boldsymbol{A})$, and $\boldsymbol{B}$ has a non-zero minor $\boldsymbol{N}$ with order $mr(\boldsymbol{A})$, if there exist $mr(\boldsymbol{A})$ entries in N make one matching, and these entries add the others entries out of $\boldsymbol{N}$ in $\boldsymbol{B}$ can make one matching, we call $\boldsymbol{A}$ has a minimal rank submatching matching structure, briefly denote as MRSM-structure. Moreover, if there exist $mr(\boldsymbol{A})$ entries in N make one matching, and these entries add the others entries out of $\boldsymbol{N}$ in $\boldsymbol{B}$ can make one principal matching, we call $\boldsymbol{A}$ has a minimal rank submatching principal matching structure, briefly denote as MRSPMS-structure.

Moreover, we think Theorem 5.4.8 is also necessity. Therefore, we have the following conjecture:

Conjecture 5.4.10 Let $\boldsymbol{A}\in Q_n$. $\boldsymbol{A}$ allows diagonalizability if and only if $\boldsymbol{A}$ has a MRSPMS-structure.

From Theorem 2.2.2.3, we can get easily the following results:

Corollary 5.4.11 If a square sign pattern $\boldsymbol{A}$ allows diagonalizability, then $\boldsymbol{A}^k$ ($k\in\boldsymbol{N}^*$ and $\boldsymbol{A}^k$ is unambiguously defined) also allows diagonalizability with the same rank.

But the converse proposition of Corollary 5.4.11 does not hold.

Lemma 5.4.12[10] Let $\boldsymbol{B}$ be a square matrix, and λ_1, λ_2, $\cdots$, λ_m be its eigenvalues, then $\boldsymbol{B}$ can be diagonalizable if and only if $q(\boldsymbol{B})=0$, where $q(t)=(t-\lambda_1)\cdots(t-\lambda_m)$, where $q(t)$ is called as minimal polynomial of $\boldsymbol{B}$.

Lemma 5.4.13[10] Let $\boldsymbol{B}$ be a $n\times n$ matrix, and $q(x)$ is minimal polynomial of $\boldsymbol{B}$, if $f(\boldsymbol{B})=0$, then $q(x)\mid f(x)$.

Theorem 5.4.14 A $n\times n$ sign pattern $\boldsymbol{A}$ allows diagonalizability if and only if there exists a real matrix $\boldsymbol{B}\in Q(\boldsymbol{A})$ and a one variable polynomials $f(x)$ such that $\boldsymbol{B}f(\boldsymbol{B})=0$ holds, where the constant term of $f(x)$ is not equal to 0.

Proof Sufficiency. In term of Lemma 5.4.13, $\boldsymbol{B}f(\boldsymbol{B})=0$, and the constant term of $f(x)$ is not equal to 0, then $q(x)\mid xf(x)$. Thus there exists the factor $x-0$ or not in $q(x)$.

Case 1. If there exists the factor $x-0$ in $q(x)$, then the times of the factor $x-0$ is one. Thus the the algebraic multiplicity and geometric multiplicity of the eigenvalue 0 of $\boldsymbol{B}$ are equal. In term of Corollary 5.4.3, $\boldsymbol{B}$ is rank-Principal. So $\boldsymbol{A}$ allows diagonalizability

Case 2. If there does not exist the factor $x-0$ in $q(x)$, then 0 is not the root of $q(x)$. Furthermore, 0 is not the root of character polynomial $p(x)$ of $\boldsymbol{B}$. So $\boldsymbol{B}$ is invertible. Therefore, $\boldsymbol{B}$ is rank-principal, and $\boldsymbol{A}$ allows diagonalizability.

Necessity. If $\boldsymbol{A}$ allows diagonalizability, then there exists a rank-principal matrix $\boldsymbol{B}$.

Case 1. If $\boldsymbol{B}$ is invertible, then character polynomial $f(\boldsymbol{B})=0$. So $\boldsymbol{B}f(\boldsymbol{B})=0$.

Case 2. If $\boldsymbol{B}$ is not invertible, then 0 is a eigenvalue of $\boldsymbol{B}$. Because $\boldsymbol{B}$ is rank-principal, So the algebraic multiplicity and geometric multiplicity of the eigenvalue 0 of $\boldsymbol{B}$ are equal. Therefore, there exists only one factor $x-0$ in $q(x)$. Thus, there exists at least one polynomial $f(x)$ with non-zero constant term such that $xf(x)=0$. If the degree of $f(x)$ in these polynomials is the lowest, then $q(x)=xf(x)$. So, the result holds.

Remark 5.4.15 Generally, in Theorem 5.4.14, we can find one polynomial $f(x)$ whose degree is no more than $n-1$, and we can also choose this polynomial as minimal polynomial of $\boldsymbol{B}$.

In theorem 5.4.14, if there exists a real matrix $\boldsymbol{B}\in Q(\boldsymbol{A})$ and a one variable polynomials $f(x)$ such that $\boldsymbol{B}f(\boldsymbol{B}) = 0$ holds, then sign pattern $\boldsymbol{A}$ allows diagonalizability. Moreover, we can get the rank of the diagonalizable matrix, which is equal to n minus the algebraic multiplicity (or geometric multiplicity) of the eigenvalue 0 of $\boldsymbol{B}$.

In [10], it presented that every sign k-potent sign pattern allows diagonalizability with rank $mr(\boldsymbol{A})$. In fact, we can get the same results with rank $mr(\boldsymbol{A})$ for every $\boldsymbol{A}^p$, $p \in \mathbf{N}^*$. So we have the following results:

Theorem 5.4.16 If $\boldsymbol{A}$ is a sign k-potent sign pattern, then $\boldsymbol{A}^p$, $p \in \mathbf{N}^*$ allows diagonalizability with rank $mr(\boldsymbol{A})$.

Proof In terms of Theorem 5.4.7, we can get $\boldsymbol{A}$ allows diagonalizability. Therefore, there exists a $\boldsymbol{B} \in Q(\boldsymbol{A})$ that can be diagonalizable, and $\boldsymbol{B}^p \in Q(\boldsymbol{A}^p)$, $p \in \mathbf{N}^*$ also is diagonalizable. So $\boldsymbol{A}^p$, $p \in \mathbf{N}^*$ allows diagonalizability.

Moreover, we can find $rank(\boldsymbol{B}^p)$ is also equal $mr(\boldsymbol{A})$. Therefore, $\boldsymbol{A}^p$, $p \in \mathbf{N}^*$ allows diagonalizability with rank $mr(\boldsymbol{A})$.

Moreover, we also can get the following results:

Theorem 5.4.17 Every sign k-potent sign pattern $\boldsymbol{A}$ allows diagonalizability with rank $MR(\boldsymbol{A})$.

Proof For sign k-potent sign pattern $\boldsymbol{A}$, we can get $mr(\boldsymbol{A}^p) = mr(\boldsymbol{A})$, $MR(\boldsymbol{A}^p) = MR(\boldsymbol{A})$, $p \in \mathbf{N}^*$. For $\boldsymbol{B}$, $\boldsymbol{C} \in Q(\boldsymbol{A})$ with $rank(\boldsymbol{B}) = rank(\boldsymbol{C}) = MR(\boldsymbol{A})$, if $rank(\boldsymbol{BC}) = MR(\boldsymbol{A})$, we will find $\boldsymbol{B}$ and $\boldsymbol{C}$ are rank-principal.

Without loss of generality, we may assume that the front $MR(\boldsymbol{A})$ columns of $\boldsymbol{B}$ are linearly independent, that is $\boldsymbol{B} = [\boldsymbol{B}_1 \quad \boldsymbol{B}_2]$. And let $\boldsymbol{C} = \begin{bmatrix} \boldsymbol{C}_1 \\ \boldsymbol{C}_2 \end{bmatrix}$. We can use suitable elementary column operations to reduce $\boldsymbol{B}$ as $[\boldsymbol{B}_1 \quad 0]$. Therefore, $\boldsymbol{BC} = [\boldsymbol{B}_1 \quad \boldsymbol{B}_2]\begin{bmatrix} \boldsymbol{C}_1 \\ \boldsymbol{C}_2 \end{bmatrix} = [\boldsymbol{B}_1 \quad 0]\begin{bmatrix} \boldsymbol{C}_1 \\ \boldsymbol{C}_2' \end{bmatrix} = \boldsymbol{B}_1\boldsymbol{C}_1$ and $rank(\boldsymbol{BC}) = rank(\boldsymbol{C}_1) = MR(\boldsymbol{A})$.

For $\boldsymbol{B}$, $\boldsymbol{C} \in Q(\boldsymbol{A})$ with $rank(\boldsymbol{B}) = rank(\boldsymbol{B}_1) = rank(\boldsymbol{C}) = rank(\boldsymbol{C}_1) = MR(\boldsymbol{A})$, we can get that matrices $\boldsymbol{B}$ and $\boldsymbol{C}$ are rank-principal, or we will get contradiction. Therefore, we prove this theorem.

5.5 Sign patterns in Frobenius normal form

The Frobenius normal form of a sign pattern $\boldsymbol{A} \in Q_n$ is a sign pattern in block upper triangular form:

$$P^{\mathrm{T}}AP=\begin{bmatrix}A_{11} & A_{12} & \cdots & A_{1p}\\ 0 & A_{22} & \cdots & A_{2p}\\ \vdots & \vdots & \vdots & \vdots\\ 0 & 0 & \cdots & A_{pp}\end{bmatrix},$$

where P is a permutation sign pattern and the diagonal blocks A_{ii} are irreducible (which are called the irreducible components of A).

By Lemma 2.1.7, permutational similarity preserves diagonalizability. So, it suffices to consider which sign patterns in Frobenius normal form allow diagonalizability.

By considering the minimal polynomials, we get the following result.

Lemma 5.5.1 If a sign pattern A in Frobenius normal form $A=\begin{bmatrix}A_{11} & A_{12} & \cdots & A_{1p}\\ 0 & A_{22} & \cdots & A_{2p}\\ \vdots & \vdots & \vdots & \vdots\\ 0 & 0 & \cdots & A_{pp}\end{bmatrix}$ allows diagonalizability, then each irreducible component A_{ii}, $1\leqslant i\leqslant p$, allows diagonalizability.

From Lemma 5.5.1, we can get its converse negative proposition.

Corollary 5.5.2 Let a sign pattern A be in Frobenius normal form $A=\begin{bmatrix}A_{11} & A_{12} & \cdots & A_{1p}\\ 0 & A_{22} & \cdots & A_{2p}\\ \vdots & \vdots & \vdots & \vdots\\ 0 & 0 & \cdots & A_{pp}\end{bmatrix}$. If there exists an irreducible component A_{ii}, $1\leqslant i\leqslant p$, that does not allow diagonalizability, then A also does not allow diagonalizability.

In fact, Lemma 5.5.1 and Corollary 5.5.2 present a new method to decide whether a matrix allows diagonalizability.

Theorem 5.5.3 A square sign pattern A in Frobenius normal form $A=\begin{bmatrix}A_{11} & A_{12} & \cdots & A_{1p}\\ 0 & A_{22} & \cdots & A_{2p}\\ \vdots & \vdots & \vdots & \vdots\\ 0 & 0 & \cdots & A_{pp}\end{bmatrix}$ allows diagonalizability if and only if there exists a real matrix $B = \begin{bmatrix} B_{11} & B_{12} & \cdots & B_{1p} \\ 0 & B_{22} & \cdots & B_{2p} \\ \vdots & \vdots & \vdots & \vdots \\ 0 & 0 & \cdots & B_{pp} \end{bmatrix} \in Q(A)$ (where $B_{ii} \in Q(A_{ii})$) such that $rank(B) = rank(B_{11}) + rank(B_{22}) + \cdots + rank(B_{pp})$, and each B_{ii} is diagonalizable.

Proof Sufficiency. Assume that $B = \begin{bmatrix} B_{11} & B_{12} & \cdots & B_{1p} \\ 0 & B_{22} & \cdots & B_{2p} \\ \vdots & \vdots & \vdots & \vdots \\ 0 & 0 & \cdots & B_{pp} \end{bmatrix} \in Q(A)$ (where $B_{ii} \in Q(A_{ii})$), $rank(B) = rank(B_{11}) + rank(B_{22}) + \cdots + rank(B_{pp})$ and each of $B_{11}, B_{22}, \cdots, B_{pp}$ is diagonalizable. By Theorem 5.2.5, each B_{ii} has a rank-principal certificate. In view of $rank(B) = rank(B_{11}) + \cdots + rank(B_{pp})$, the smallest principal submatrix containing all these certificates forms a rank-principal certificate of B. By Theorem 5.2.5, A allows diagonalizability.

Necessity. Assume that A allows diagonalizability. Let $B = \begin{bmatrix} B_{11} & B_{12} & \cdots & B_{1p} \\ 0 & B_{22} & \cdots & B_{2p} \\ \vdots & \vdots & \vdots & \vdots \\ 0 & 0 & \cdots & B_{pp} \end{bmatrix} \in Q(A)$ be a diagonalizable matrix. Then the minimal polynomial of B has no repeated roots, and thus the same holds for each B_{ii}. Hence, each B_{ii} is diagonalizable. Further, $rank(B)$ is equal to the number of non-zero eigenvalues of B, and hence, $rank(B) = rank(B_{11}) + \cdots + rank(B_{pp})$.

Corollary 5.5.4 If a sign pattern in Frobenius normal form $A = \begin{bmatrix} A_{11} & A_{12} & \cdots & A_{1p} \\ 0 & A_{22} & \cdots & A_{2p} \\ \vdots & \vdots & \vdots & \vdots \\ 0 & 0 & \cdots & A_{pp} \end{bmatrix}$ allows diagonalizability, then the set of the ranks of diagonalizable matrices in $Q(A)$, is a subset of the set of ranks of diagonalizable matrices in the qualitative class of the block diagonal sign pattern

$$\begin{bmatrix} A_{11} & 0 & \cdots & 0 \\ 0 & A_{22} & \cdots & 0 \\ \vdots & \vdots & \vdots & \vdots \\ 0 & 0 & \cdots & A_{pp} \end{bmatrix}.$$

Corollary 5.5.5 If $mr(A) \geqslant MR(\Lambda)$, then A does not allow diagonalizability,

where $\Lambda = \begin{bmatrix} A_{11} & 0 & \cdots & 0 \\ 0 & A_{22} & \cdots & 0 \\ \vdots & \vdots & \vdots & \vdots \\ 0 & 0 & \cdots & A_{pp} \end{bmatrix}$.

The following fact is useful when studying sign patterns in block form that allow diagonalizability.

Lemma 5.5.6 Let $A = \begin{bmatrix} A_{11} & A_{12} \\ A_{21} & A_{22} \end{bmatrix} \in \mathrm{Q}_n$. If $mr(A) = mr(A_{11})$, then there exists a real matrix $B = \begin{bmatrix} B_{11} & B_{12} \\ B_{21} & B_{22} \end{bmatrix} \in \mathrm{Q}(A)$ and a non-singular submatrix C of B_{11} such that $rank(C) = rank(B_{11}) = rank(B) = \mathrm{mr}(A)$.

Proof Pick a matrix $B = \begin{bmatrix} B_{11} & B_{12} \\ B_{21} & B_{22} \end{bmatrix} \in Q(A)$ with $rank(B) = mr(A)$, where each $B_{ij} \in Q(A_{ij})$. Then $rank(B_{11}) \leqslant rank(B) = mr(A) = mr(A_{11})$. But of course we also have the opposite inequality $rank(B_{11}) \geqslant mr(A_{11})$. It follows that $rank(B_{11}) = mr(A_{11}) = mr(A)$. Thus B_{11} has a non-singular submatrix C of rank $mr(A)$.

In [9], two related open problems Problem 5.5 and 5.6 is presented. We phrase an interesting open combinatorial sufficient condition for a symmetrically partitioned block upper triangular sign pattern to allow diagonalizability.

Problem 5.5.7 Let A be a sign pattern in symmetrically partitioned block upper triangular form $A = \begin{bmatrix} A_{11} & A_{12} & \cdots & A_{1p} \\ 0 & A_{22} & \cdots & A_{2p} \\ \vdots & \vdots & \vdots & \vdots \\ 0 & 0 & \cdots & A_{pp} \end{bmatrix}$. Suppose that for each $i = 1, 2, \cdots, p$, $mr(A_{ii}) = mr((A_{ii}, A_{i(i+1)} \cdots A_{ip}))$ and each A_{ii} allows diagonalizability. Does it then necessarily follow that A allows diagonalizability?

A related open problem is the following.

Problem 5. 5. 8 Let A_1 be a square sign pattern that allows rank-principality. Is it true that for every sign pattern $A = [A_1 \quad A_2]$ such that $mr(A_1) = mr(A)$, A allows rank-principality?

We note that an affirmative answer to Problem 5. 5. 8 implies an affirmative answer to Problem 5. 5. 7.

Next we can solve partly out these problems. We may add some conditions to get a sufficient condition.

Theorem 5. 5. 9 Let A_1 be a square sign pattern that allows rank-principality, and there is the submatrix A_1^0 of sign patterns of one rank-principal certificate C of A_1, M_0 which support A_1^0 is a submatching of M_1 which support C, if $mr(A_1) = mr(A_1^0) = mr(A)$, then $A = [A_1 \quad A_2]$ allows rank-principality.

Proof The proof is similar to Theorem 5. 3. 9 and 5. 4. 8, in terms of Lemma 2. 2. 2. 3, we can find one submatrix B_0 of $B \in Q(A)$ with $rank(B) = mr(A)$, and $sgn(B_0) = M_0$, through every time we only change one entries of submatrix B_1 in B, and $sgn(B_1) = M_1$, we can get a series of new matrices whose rank gap is 1. Therefore, we can make order of C times changes to matrix B, and we will get a new matrix $B' \in Q(A)$, whose rank is order of C, and the new matrix B' is rank-principal.

According to Theorem 5. 5. 9, we can also get a sufficient condition of Problem 5. 6.

Theorem 5. 5. 10 Let A be a sign pattern in symmetrically partitioned block upper triangular form $A = \begin{bmatrix} A_{11} & A_{12} & \cdots & A_{1p} \\ 0 & A_{22} & \cdots & A_{2p} \\ \vdots & \vdots & \vdots & \vdots \\ 0 & 0 & \cdots & A_{pp} \end{bmatrix}$. Suppose that for each $i = 1, 2, \cdots, p$, $mr(A_{ii}) = mr((A_{ii}, A_{i(i+1)} \quad \cdots \quad A_{ip}))$ and each A_{ii} allow diagonalizability, and has a MRSPMS-structure, then A allows diagonalizability.

In fact, we can improve the conditions of Theorem 5. 5. 10, so we have the following theorem:

Theorem 5. 5. 11 Let A be a sign pattern in symmetrically partitioned block upper triangular form $\boldsymbol{A} = \begin{bmatrix} \boldsymbol{A}_{11} & \boldsymbol{A}_{12} & \cdots & \boldsymbol{A}_{1p} \\ 0 & \boldsymbol{A}_{22} & \cdots & \boldsymbol{A}_{2p} \\ \vdots & \vdots & \vdots & \vdots \\ 0 & 0 & \cdots & \boldsymbol{A}_{pp} \end{bmatrix}$. If

$mr(\boldsymbol{A}) = mr(\boldsymbol{\Lambda}) = mr(\begin{bmatrix} \boldsymbol{A}_{11} & 0 & \cdots & 0 \\ 0 & \boldsymbol{A}_{22} & \cdots & 0 \\ \vdots & \vdots & \vdots & \vdots \\ 0 & 0 & \cdots & \boldsymbol{A}_{pp} \end{bmatrix})$, each $\boldsymbol{A}_{ii}$ allows diagonalizability and has a MRSPMS-structure, then $\boldsymbol{A}$ allows diagonalizability.

Chapter 6

Some Attempts of Sign Patterns That Allow Diagonalizability from Different Direction

The eigen-problem is an important research field in both the tradition and sign pattern matrix, and this often establish relationships with the diagonalizability of matrix. In this chapter, we mainly consider sign patterns that allow simultaneously unitary diagonalizability. Moreover, in this chapter we also obtain some results of 1 ~4 order sign patterns that allow diagonalizability. Through this way we want to find its rules and finally to solve this problem.

6.1 Sign patterns allowing simultaneous unitary diagonalizability

6.1.1 Allowing simultaneously unitary diagonalizability of sign patterns

In this section, we consider two sign patterns allowing simultaneous unitary diagonalizability.

Definition 6.1.1 Let $\boldsymbol{A} \in Q_n$. If there exists a real matrix $\boldsymbol{B} \in Q(\boldsymbol{A})$ such that $\boldsymbol{B}$ has property $\boldsymbol{B}\boldsymbol{B}^{\mathrm{T}} = \boldsymbol{B}^{\mathrm{T}}\boldsymbol{B}$, then we say $\boldsymbol{A}$ allows unitary diagonalizability.

Lemma 6.1.2 $\boldsymbol{A}$, $\boldsymbol{B} \in Q_n$ are sign patterns allowing simultaneous unitary diagonalizability if and only if there exist $\boldsymbol{A}_0 \in Q(\boldsymbol{A})$, $\boldsymbol{B}_0 \in Q(\boldsymbol{B})$, such that $\boldsymbol{A}_0$ and $\boldsymbol{B}_0$ are simultaneous diagonalizable and $\boldsymbol{A}_0\boldsymbol{B}_0 = \boldsymbol{B}_0\boldsymbol{A}_0$.

Proof $\boldsymbol{A}$, $\boldsymbol{B} \in Q_n$ are sign patterns allowing simultaneous diagonalizability if and only if there exist $\boldsymbol{A}_0 \in Q(\boldsymbol{A})$, $\boldsymbol{B}_0 \in Q(\boldsymbol{B})$, such that $\boldsymbol{A}_0$ and $\boldsymbol{B}_0$ are simultaneous diagonalizable. This holds if and only if $\boldsymbol{A}_0\boldsymbol{B}_0 = \boldsymbol{B}_0\boldsymbol{A}_0$.

Theorem 6.1.3 If $\boldsymbol{A}$ and $\boldsymbol{B}$ are two non-negative sign patterns allowing simultaneous unitary diagonalizability, then $\boldsymbol{AB} = \boldsymbol{BA}$.

Proof By Lemma 6.1.2, $\boldsymbol{A}$, $\boldsymbol{B} \in Q_n$ are two sign patterns allowing simultaneous diagonalizability if and only if there exist $\boldsymbol{A}_0 \in Q(\boldsymbol{A})$, $\boldsymbol{B}_0 \in Q(\boldsymbol{B})$, such that $\boldsymbol{A}_0\boldsymbol{B}_0 = \boldsymbol{B}_0\boldsymbol{A}_0$. Because $\boldsymbol{A}$ and $\boldsymbol{B}$ are non-negative, the proof is similar to that of Lemma 6.1.2. If $(\boldsymbol{A}_0\boldsymbol{B}_0)_{ij} = 0$ $(i, j = 1, 2, \cdots, n)$, then

$$(\boldsymbol{AB})_{ij} = 0.$$

Likewise, if $(\boldsymbol{A}_0\boldsymbol{B}_0)_{\mathrm{ij}} > 0$, then

$$(\boldsymbol{AB})_{ij} = +.$$

Vice versa, if $(B_0A_0)_{ij}=0$, then $(BA)_{ij}=0$, and if $(B_0A_0)_{ij}>0$, then $(BA)_{ij}=+$. Therefore, according to $A_0B_0=B_0A_0$, $AB=BA$ holds.

Similarly, we can easily obtain the following result:

Corollary 6.1.4 Let A, $B\in Q_n$ be sign patterns allowing simultaneous diagonalizability. $(AB)_{ij}=\#$ if and only if $(BA)_{ij}=\#$, i, $j=1, 2, \cdots, n$, then $AB=BA$.

Lemma 6.1.5 [2, Corollary 3.7] Let A, $B\in M_n$ be two non-singular Hermitian matrices simultaneously unitary diagonalizable. Then, there is a Hermitian matrix $X\in M_n$ such that $B=XAX$ if and only if there is a unitary matrix $V\in M_n$ such that V^*A_0V and V^*B_0V are diagonal matrices of the forms

$$V^*A_0V=S_A\oplus A_1\oplus\cdots\oplus A_1,\quad V^*B_0V=S_B\oplus B_1\oplus\cdots\oplus B_1,$$

where $\mathrm{sgn}(S_A)=\mathrm{sgn}(S_B)$, and A_i, $B_i\in M_2$ are indefinite matrices such that B_i is a negative multiple of A_i^{-1}, $i=1, 2, \cdots, 1$.

By Lemma 6.1.5, we can easily obtain the following theorem:

Theorem 6.1.6 Let A, $B\in Q_n$ be two symmetric sign patterns allowing simultaneous unitary diagonalizability and $MR(A)=MR(B)=n$. Then, there exist a symmetric matrix $X\in M_n$ and non-singular $A_0\in Q(A)$, $B_0\in Q(B)$, such that $B_0=XA_0X$ if and only if there is an orthogonal matrix $V\in M_n$ such that V^*A_0V and V^*B_0V are diagonal matrices of the forms

$$V^*A_0V=S_A\oplus A_1\oplus\cdots\oplus A_1,\quad V^*B_0V=S_B\oplus B_1\oplus\cdots\oplus B_1,$$

where $\mathrm{sgn}(S_A)=\mathrm{sgn}(S_B)$, and A_i, $B_i\in M_2$ are indefinite matrices such that B_i is a negative multiple of A_i^{-1}, $i=1, 2, \cdots, l$.

Theorem 6.1.7 Let A, $B\in Q_n$ be two non-negative symmetric sign patterns allowing simultaneous unitary diagonalizability. If there are non-singular $A_0\in Q(A)$, $B_0\in Q(B)$, and a non-negative real symmetric matrix X_0 such that $B_0=X_0A_0X_0$, then there exists a symmetric sign pattern matrix X such that $B=XAX$.

Proof This theorem can be proved by using similar methods of Theorem 6.1.3.

Corollary 6.1.8 Let A, $B\in Q_n$ be two symmetric sign patterns allowing simultaneous unitary diagonalizability. If there are $A_0\in Q(A)$, $B_0\in Q(B)$, and a non-negative real symmetric matrix X_0 such that $B_0=X_0A_0X_0$, and there is not # in product of $S(X_0)\ AS(X_0)$, then there exists a symmetric sign pattern matrix $X=$

$S(X_0)$ such that $B = XAX$.

Proof If there is not # in product of $S(X_0)AS(X_0)$, by $B_0 = X_0A_0X_0$, we have

$$\operatorname{sgn}((X_0A_0X_0)_{ij}) = (XAX)_{ij}, \text{ for all } i, j = 1, 2, \cdots, n.$$

Moreover, $\operatorname{sgn}((B_0)_{ij}) = (B)_{ij}$, for all i, $j = 1, \cdots, n$. Thus $B = XAX$ holds

Lemma 6.1.9 Let A and B be two $n \times n$ non-singular simultaneous diagonalizable normal real matrices. Let the eigenvalues of A be $a_1, \cdots, a_k, \alpha_{k+1} + i\beta_{k+1}, \cdots, \alpha_p + i\beta_p$, and the eigenvalues of B be $b_1, \cdots, b_k, \gamma_{k+1} + i\omega_{k+1}, \cdots, \gamma_p + i\omega_p$ If

$$\begin{cases} a_ib_i = a_jb_j & i, j = 1, \cdots, k, \\ \alpha_i\omega_j = \beta_i\gamma_j & i, j = k+1, \cdots, p, \end{cases}$$

then there exists a nonsingular symmetric matrix X such that $B = XAX$.

Proof Let A and B be two non-singular simultaneous diagonalizable normal real matrices, and there exists real orthogonal matrix Q such that

$$Q^{T}AQ = \begin{bmatrix} a_1 & & & & & & & 0 \\ & \ddots & & & & & & \\ & & a_k & & & & & \\ & & & \alpha_{k+1} & \beta_{k+1} & & & \\ & & & -\beta_{k+1} & \alpha_{k+1} & & & \\ & & & & & \ddots & & \\ & & & & & & \alpha_p & \beta_p \\ 0 & & & & & & -\beta_p & \alpha_p \end{bmatrix}$$

and

$$Q^{T}BQ = \begin{bmatrix} b_1 & & & & & & & 0 \\ & \ddots & & & & & & \\ & & b_k & & & & & \\ & & & \gamma_{k+1} & \omega_{k+1} & & & \\ & & & -\omega_{k+1} & \gamma_{k+1} & & & \\ & & & & & \ddots & & \\ & & & & & & \gamma_p & \omega_p \\ 0 & & & & & & -\omega_p & \gamma_p \end{bmatrix}.$$

Suppose that there exist a non-singular symmetric matrix X such that $Q^{T}AQ = XQ^{T}BQX$, then

$$X^{-1} = \begin{bmatrix} a_1 & & & & & & & & 0 \\ & \ddots & & & & & & & \\ & & a_k & & & & & & \\ & & & \alpha_{k+1} & \beta_{k+1} & & & & \\ & & & -\beta_{k+1} & \alpha_{k+1} & & & & \\ & & & & & \ddots & & & \\ & & & & & & & \alpha_p & \beta_p \\ 0 & & & & & & & -\beta_p & \alpha_p \end{bmatrix}^{-1}$$

$$X\begin{bmatrix} b_1 & & & & & & & & 0 \\ & \ddots & & & & & & & \\ & & b_k & & & & & & \\ & & & \gamma_{k+1} & \omega_{k+1} & & & & \\ & & & -\omega_{k+1} & \gamma_{k+1} & & & & \\ & & & & & \ddots & & & \\ & & & & & & & \gamma_p & \omega_p \\ 0 & & & & & & & -\omega_p & \gamma_p \end{bmatrix}.$$

We divide the three matrices into 2 × 2 blocks with the suitable dimension. Then, their product will have the following three kinds of equations.

Case 1:

$$\begin{bmatrix} a_1 & \cdots & 0 \\ \vdots & \vdots & \vdots \\ 0 & \cdots & a_k \end{bmatrix}^{-1} X_{11} \begin{bmatrix} b_1 & \cdots & 0 \\ \vdots & \vdots & \vdots \\ 0 & \cdots & b_k \end{bmatrix} = \begin{bmatrix} b_1 & \cdots & 0 \\ \vdots & \vdots & \vdots \\ 0 & \cdots & b_k \end{bmatrix}^{T} X_{11} \left[\begin{bmatrix} a_1 & \cdots & 0 \\ \vdots & \vdots & \vdots \\ 0 & \cdots & a_k \end{bmatrix}^{-1}\right]^{T},$$

where $X = \begin{bmatrix} X_{11} & X_{12} \\ X_{21} & X_{22} \end{bmatrix}$ and X_{11} is a symmetric square matrix.

From above equation, we find that, only need let $a_i b_i = a_j b_j$, $i, j = 1, 2, \cdots, k$, the above equation constantly holds. Thus, there exists solution X_{11}.

Case 2:

$$\begin{bmatrix} a_1 & \cdots & 0 \\ \vdots & \ddots & \vdots \\ 0 & \cdots & a_k \end{bmatrix}^{-1} X_{12} \begin{bmatrix} \gamma_{k+1} & \omega_{k+1} & \cdots & 0 & 0 \\ -\omega_{k+1} & \gamma_{k+1} & \cdots & 0 & 0 \\ \cdots & \cdots & \cdots & \cdots & \cdots \\ 0 & 0 & \cdots & \gamma_p & \omega_p \\ 0 & 0 & \cdots & -\omega_p & \gamma_p \end{bmatrix} = \begin{bmatrix} b_1 & \cdots & 0 \\ \vdots & \ddots & \vdots \\ 0 & \cdots & b_k \end{bmatrix}^{\mathrm{T}}$$

$$X_{12} \left[\begin{bmatrix} \alpha_{k+1} & \beta_{k+1} & \cdots & 0 & 0 \\ -\beta_{k+1} & \alpha_{k+1} & \cdots & 0 & 0 \\ \cdots & \cdots & \cdots & \cdots & \cdots \\ 0 & 0 & \cdots & \alpha_p & \beta_p \\ 0 & 0 & \cdots & -\beta_p & \alpha_p \end{bmatrix}^{-1} \right]^{\mathrm{T}}$$

By $a_i b_i = a_j b_j$, we have

$$X_{12} \begin{bmatrix} \gamma_{k+1} & \omega_{k+1} & \cdots & 0 & 0 \\ -\omega_{k+1} & \gamma_{k+1} & \cdots & 0 & 0 \\ \cdots & \cdots & \cdots & \cdots & \cdots \\ 0 & 0 & \cdots & \gamma_p & \omega_p \\ 0 & 0 & \cdots & -\omega_p & \gamma_p \end{bmatrix} \left[\begin{bmatrix} \alpha_{k+1} & \beta_{k+1} & \cdots & 0 & 0 \\ -\beta_{k+1} & \alpha_{k+1} & \cdots & 0 & 0 \\ \cdots & \cdots & \cdots & \cdots & \cdots \\ 0 & 0 & \cdots & \alpha_p & \beta_p \\ 0 & 0 & \cdots & -\beta_p & \alpha_p \end{bmatrix} \right]^{\mathrm{T}} = a_i b_i X_{12}.$$

Because $\alpha_l \pm i\beta_l$ *and* $\gamma_l \pm i\omega_l$ are imaginary characteristic root of $\boldsymbol{A}$ and $\boldsymbol{B}$, X_{12} has a unique solution, $l = k+1,\ k+2,\ \cdots,\ p$, and $X_{21} = X_{12}^{\mathrm{T}}$.

Case 3:

$$\begin{bmatrix} \alpha_{k+1} & \beta_{k+1} & \cdots & 0 & 0 \\ -\beta_{k+1} & \alpha_{k+1} & \cdots & 0 & 0 \\ \cdots & \cdots & \cdots & \cdots & \cdots \\ 0 & 0 & \cdots & \alpha_p & \beta_p \\ 0 & 0 & \cdots & -\beta_p & \alpha_p \end{bmatrix}^{-1} X_{22} \begin{bmatrix} \gamma_{k+1} & \omega_{k+1} & \cdots & 0 & 0 \\ -\omega_{k+1} & \gamma_{k+1} & \cdots & 0 & 0 \\ \cdots & \cdots & \cdots & \cdots & \cdots \\ 0 & 0 & \cdots & \gamma_p & \omega_p \\ 0 & 0 & \cdots & -\omega_p & \gamma_p \end{bmatrix} =$$

$$\begin{bmatrix} \gamma_{k+1} & \omega_{k+1} & \cdots & 0 & 0 \\ -\omega_{k+1} & \gamma_{k+1} & \cdots & 0 & 0 \\ \cdots & \cdots & \cdots & \cdots & \cdots \\ 0 & 0 & \cdots & \gamma_p & \omega_p \\ 0 & 0 & \cdots & -\omega_p & \gamma_p \end{bmatrix}^{\mathrm{T}} X_{22} \left[\begin{bmatrix} \alpha_{k+1} & \beta_{k+1} & \cdots & 0 & 0 \\ -\beta_{k+1} & \alpha_{k+1} & \cdots & 0 & 0 \\ \cdots & \cdots & \cdots & \cdots & \cdots \\ 0 & 0 & \cdots & \alpha_p & \beta_p \\ 0 & 0 & \cdots & -\beta_p & \alpha_p \end{bmatrix}^{-1} \right]^{\mathrm{T}}$$

Unfold this equation, then $\alpha_{k+s-1}\omega_{k+s-1} = \gamma_{k+s-1}\beta_{k+s-1}$ $(1 \leqslant s \leqslant n-k+1)$

can make that the above equation has solution X_{12}.

According to above analysis, the proof is completed.

Theorem 6.1.10 Let A, $B \in Q_n$ be two non-negative sign patterns allowing simultaneous unitary diagonalizability. And there is $A_0 \in Q(A)$, $B_0 \in Q(B)$ such that A_0 and B_0 are two $n \times n$ non-singular simultaneously diagonalizable normal real matrices. Let the eigenvalues of A_0 be a_1, a_2, $\cdots$, a_k, $\alpha_{k+1} + i\beta_{k+1}$, $\alpha_{k+2} + i\beta_{k+2}$, $\cdots$, $\alpha_p + i\beta_p$ and the eigenvalues of B_0 be b_1, b_2, $\cdots$, b_k, $\gamma_{k+1} + i\omega_{k+1}$, $\gamma_{k+2} + i\omega_{k+2}$, $\cdots$, $\gamma_p + i\omega_p$, and

$$\begin{cases} a_i b_i = a_j b_j & i, j = 1, 2, \cdots, k, \\ \alpha_i \omega_j = \beta_i \gamma_j & i, j = k+1, k+2, \cdots, p, \end{cases}$$

then there exists a symmetric sign pattern X such that $B = XAX$ if and only if there does not exist # entries in XAX.

Proof By Lemma 6.1.9, we know that there exists X_0 such that $B_0 = X_0 A_0 X_0$. Let $X = X_0$. Because A and B are non-negative sign patterns, similar to Corollary 6.1.8, we can also obtain that $B = XAX$ holds if and only if there does not exist # entries in XAX.

Corollary 6.1.11 Let A, $B \in Q_n$ be sign patterns allowing simultaneous unitary diagonalizability. If there are $A_0 \in Q(A)$, $B_0 \in Q(B)$ and a real symmetric matrix X_0 such that $B_0 = X_0 A_0 X_0$, and there is not # in product of $S(X_0)$ $AS(X_0)$, then there exists a symmetric sign pattern matrix $X = S(X_0)$ such that $B = XAX$.

6.1.2 Conclusion

In this part, we make a discussion about allowing unitary diagonalizability of sign pattern. Some sufficient and necessary conditions of allowing unitary diagonalizability are also obtained. Moreover, the relation of two sign patterns allowing simultaneous unitary diagonalizability is researched.

6.2 Up to 4 order sign patterns that allow diagonalizability

Lemma 6.2.1 [8] Let A be an [7] matrix. Then A will be permutationally similar to a $k \times k$ lower triangular block matrix (also can be upper triangular block matrix)

$$\begin{bmatrix} A_1 & \cdots & 0 & 0 & \cdots & 0 \\ \vdots & \vdots & \vdots & \vdots & \vdots & \vdots \\ 0 & \cdots & A_g & 0 & \cdots & 0 \\ A_{g+1,1} & \cdots & A_{g+1,g} & A_{g+1} & \cdots & 0 \\ \vdots & \vdots & \vdots & \vdots & \vdots & \vdots \\ A_{k,1} & \cdots & A_{k,g} & A_{k,g+1} & \cdots\cdots & A_k \end{bmatrix},$$

where A_1, $\cdots A_g$, A_{g+1}, $\cdots$, A_k are irreducible square matrices, $g < q \leqslant k$, and not all $A_{q,1}$, $\cdots A_{q,q-1}$ are 0. These kinds of matrices are also collectively referred to as the Frobenius normal form.

If a sign pattern is 1 order square matrix, then it allows diagonalizability. For a 2 order square sign pattern, we have the following results:

Theorem 6.2.2 Let $A \in Q_2$. Only either $A = \begin{bmatrix} 0 & a_{12} \\ 0 & 0 \end{bmatrix}$ or $A = \begin{bmatrix} 0 & 0 \\ a_{21} & 0 \end{bmatrix}$ does not allow diagonalizability. In contrast, all other 2-order square sign patterns allow diagonalizability, where

$$a_{21} a_{12} \neq 0.$$

Proof If sign pattern A only has a non-zero diagonal entry, then $c(A) \geqslant 1$. In terms of Theorem 2.1.5, we can get that A allows diagonalizability. If $a_{21} a_{12} \neq 0$, then $D(A)$ owns a composite cycle with length 2, and $c(A) = 2$. Therefore, in terms of Theorem 2.1.5, we can also get that A allows diagonalizability.

For sign pattern $A = \begin{bmatrix} 0 & a_{12} \\ 0 & 0 \end{bmatrix}$ $(a_{12} \neq 0)$, we can obtain $\forall B \in Q(A)$, with $rank(B) = 1$. No 1-order non-zero principal minor (nor a non-singular principal submatrix) exists. In terms of Theorem 2.2.2.3, we can get that A does not allow diagonalizability. Similarly, we can obtain that $A = \begin{bmatrix} 0 & 0 \\ a_{21} & 0 \end{bmatrix}$ also does not allow diagonalizability. Therefore, the result holds.[7]

Remark 6.2.3 In Theorem 6.2.2, both a_{12} and a_{21} are equal to + or −. Similarly, the entries of sign patterns in the following theorems are similar to a_{12} and a_{21}.

For a 3-order square sign pattern, the following results are obtained:

Theorem 6.2.4 Let $A \in Q_3$. A allows diagonalizability, except A is equal to the following sign patterns:

(1) All the diagonal entries are equal to 0.

(1a) Only one non-zero entry is present in sign pattern $A = \begin{bmatrix} 0 & a_{12} & 0 \\ 0 & 0 & 0 \\ 0 & 0 & 0 \end{bmatrix}$,

$\begin{bmatrix} 0 & 0 & a_{13} \\ 0 & 0 & 0 \\ 0 & 0 & 0 \end{bmatrix}$, $\begin{bmatrix} 0 & 0 & 0 \\ 0 & 0 & a_{23} \\ 0 & 0 & 0 \end{bmatrix}$ and their transpose matrices, which means that the following sign patterns exist:

$$\begin{bmatrix} 0 & 0 & 0 \\ a_{12} & 0 & 0 \\ 0 & 0 & 0 \end{bmatrix}, \begin{bmatrix} 0 & 0 & 0 \\ 0 & 0 & 0 \\ a_{13} & 0 & 0 \end{bmatrix}, \begin{bmatrix} 0 & 0 & 0 \\ 0 & 0 & 0 \\ 0 & a_{23} & 0 \end{bmatrix}.$$

The following formulation for the transpose matrices of sign pattern A is similar:

(1b) Only two non-zero entries are present in sign pattern A.

$\begin{bmatrix} 0 & a_{12} & a_{13} \\ 0 & 0 & 0 \\ 0 & 0 & 0 \end{bmatrix}$, $\begin{bmatrix} 0 & a_{12} & 0 \\ 0 & 0 & a_{23} \\ 0 & 0 & 0 \end{bmatrix}$, $\begin{bmatrix} 0 & 0 & a_{13} \\ 0 & 0 & a_{23} \\ 0 & 0 & 0 \end{bmatrix}$, $\begin{bmatrix} 0 & a_{12} & 0 \\ 0 & 0 & 0 \\ a_{31} & 0 & 0 \end{bmatrix}$, $\begin{bmatrix} 0 & a_{12} & 0 \\ 0 & 0 & 0 \\ 0 & a_{32} & 0 \end{bmatrix}$,

$\begin{bmatrix} 0 & 0 & a_{13} \\ 0 & 0 & 0 \\ 0 & a_{32} & 0 \end{bmatrix}$ and their transpose matrices.

(1c) Only three non-zero entries are present in sign pattern A

$\begin{bmatrix} 0 & a_{12} & a_{13} \\ 0 & 0 & a_{23} \\ 0 & 0 & 0 \end{bmatrix}$, $\begin{bmatrix} 0 & a_{12} & a_{13} \\ 0 & 0 & 0 \\ 0 & a_{32} & 0 \end{bmatrix}$, $\begin{bmatrix} 0 & 0 & a_{13} \\ a_{21} & 0 & a_{23} \\ 0 & 0 & 0 \end{bmatrix}$ and their transpose matrices.

(2) Only one non-zero diagonal entry is present in sign pattern A.

(2a) Only one non-zero entry exists, except for the non-zero diagonal entry in sign pattern A

$\begin{bmatrix} a_{11} & 0 & 0 \\ 0 & 0 & a_{23} \\ 0 & 0 & 0 \end{bmatrix}$, $\begin{bmatrix} 0 & 0 & 0 \\ 0 & a_{22} & 0 \\ a_{31} & 0 & 0 \end{bmatrix}$, $\begin{bmatrix} 0 & 0 & 0 \\ a_{21} & 0 & 0 \\ 0 & 0 & a_{33} \end{bmatrix}$ and their transpose matrices.

(2b) Only two non-zero entries exist, except for the non-zero diagonal entry in sign pattern $\boldsymbol{A}$

$$\begin{bmatrix} a_{11} & a_{12} & 0 \\ 0 & 0 & a_{23} \\ 0 & 0 & 0 \end{bmatrix}, \begin{bmatrix} a_{11} & a_{12} & 0 \\ 0 & 0 & 0 \\ 0 & a_{32} & 0 \end{bmatrix}, \begin{bmatrix} a_{11} & a_{12} & 0 \\ 0 & 0 & 0 \\ a_{31} & 0 & 0 \end{bmatrix}, \begin{bmatrix} a_{11} & 0 & a_{13} \\ 0 & 0 & a_{23} \\ 0 & 0 & 0 \end{bmatrix},$$

$$\begin{bmatrix} a_{11} & 0 & a_{13} \\ 0 & 0 & 0 \\ 0 & a_{32} & 0 \end{bmatrix}, \begin{bmatrix} 0 & a_{12} & a_{13} \\ 0 & a_{22} & 0 \\ 0 & 0 & 0 \end{bmatrix}, \begin{bmatrix} 0 & a_{12} & 0 \\ 0 & a_{22} & a_{23} \\ 0 & 0 & 0 \end{bmatrix}, \begin{bmatrix} 0 & a_{12} & 0 \\ 0 & a_{22} & 0 \\ a_{31} & 0 & 0 \end{bmatrix}, \begin{bmatrix} 0 & 0 & a_{13} \\ 0 & a_{22} & a_{23} \\ 0 & 0 & 0 \end{bmatrix},$$

$$\begin{bmatrix} 0 & 0 & a_{13} \\ 0 & a_{22} & 0 \\ 0 & a_{32} & 0 \end{bmatrix}, \begin{bmatrix} 0 & a_{12} & a_{13} \\ 0 & 0 & 0 \\ 0 & 0 & a_{33} \end{bmatrix}, \begin{bmatrix} 0 & a_{12} & 0 \\ 0 & 0 & a_{23} \\ 0 & 0 & a_{33} \end{bmatrix}, \begin{bmatrix} 0 & a_{12} & 0 \\ 0 & 0 & 0 \\ 0 & a_{32} & a_{33} \end{bmatrix}, \begin{bmatrix} 0 & 0 & 0 \\ 0 & a_{12} & 0 \\ a_{31} & 0 & a_{33} \end{bmatrix},$$

$$\begin{bmatrix} 0 & 0 & a_{13} \\ 0 & 0 & 0 \\ 0 & a_{32} & a_{33} \end{bmatrix}$$, and their transpose matrices.

(2c) Only three non-zero entries exist, except for the non-zero diagonal entry in sign pattern $\boldsymbol{A}$

$$\begin{bmatrix} a_{11} & a_{12} & a_{13} \\ 0 & 0 & a_{23} \\ 0 & 0 & 0 \end{bmatrix}, \begin{bmatrix} a_{11} & a_{12} & a_{13} \\ 0 & 0 & 0 \\ 0 & a_{32} & 0 \end{bmatrix}, \begin{bmatrix} 0 & a_{12} & a_{13} \\ 0 & a_{22} & a_{23} \\ 0 & 0 & 0 \end{bmatrix}, \begin{bmatrix} 0 & a_{12} & a_{13} \\ 0 & a_{22} & 0 \\ 0 & a_{32} & 0 \end{bmatrix},$$

$$\begin{bmatrix} 0 & 0 & a_{13} \\ a_{21} & a_{22} & a_{23} \\ 0 & 0 & 0 \end{bmatrix}, \begin{bmatrix} 0 & a_{12} & a_{13} \\ 0 & 0 & a_{23} \\ 0 & 0 & a_{33} \end{bmatrix}, \begin{bmatrix} 0 & a_{12} & 0 \\ 0 & 0 & 0 \\ a_{31} & a_{32} & a_{33} \end{bmatrix}, \text{ plus } \begin{bmatrix} a_{11} & a_{12} & 0 \\ 0 & 0 & 0 \\ a_{31} & a_{32} & 0 \end{bmatrix},$$

$$\begin{bmatrix} 0 & a_{12} & a_{13} \\ 0 & 0 & 0 \\ 0 & a_{32} & a_{33} \end{bmatrix}$$ with $\boldsymbol{A} = 2$, and their transpose matrices.

Proof In terms of Theorem 2. 1. 5 and Theorem 2. 1. 6, if $\boldsymbol{A} \in Q_3$ has either two diagonal entries or a composite cycle with length 3, then $\boldsymbol{A}$ allows diagonalizability. Thus, we only need to consider if the sign pattern that has at most a non-zero diagonal entry allows diagonalizability. Again according to Theorem 2. 1. 5, we only need to consider the following 6 types of sign patterns that allow diagonalizability:

(1a) In these 6 kinds of matrices, every matrix only has a non-zero entry, and their rank are 1. But in all of them there do not exist any 1 order non-zero principal minor. So these 6 kinds of matrices do not allow diagonalizability by Theorem 2.2.2.3.

(1b) In these 12 kinds of matrices, every matrix only has 2 non-zero entries, and their ranks are 1 or 2. But in all of them there do not exist any 1 or 2 order non-zero principal minor. So these 12 kinds of matrices also do not allow diagonalizability again by Theorem 2.2.2.3.

(1c) In these 6 kinds of matrices, every matrix only has 3 non-zero entries, and their rank are 2. But in all of them there do not exist any 2 order non-zero principal minor. So, these 6 kinds of matrices do not allow diagonalizability by Theorem 2.2.2.3.

(2a) In these 6 kinds of matrices, every matrix only has 2 non-zero entries, and their rank are 2. But in all of them there do not exist any 2 order non-zero principal minor. So these 6 kinds of matrices do not-allow diagonalizability by Theorem 2.2.2.3.

(2b) In these 30 kinds of matrices, every matrix only has 3 non-zero entries, and their rank are 2. But in all of them there do not exist any 2 order non-zero principal minor. So these 30 kinds of matrices do not-allow diagonalizability again by Theorem 2.2.2.3.

(2c) In the front 14 kinds of matrices, every matrix only has 4 non-zero entries, and their rank are 2. But in all of them there do not exist any 2 order non-zero principal minor. So these 14 kinds of matrices do notallow diagonalizability by Theorem 2.2.2.3. In the final 4 kinds of matrices, if their minimum rank are 1, then they have 1 order non-zero principal minor and allow diagonalizability. So they do not allow diagonalizability with minimum rank 2.

Even when a sign pattern $\boldsymbol{A} \in Q_3$ only has a diagonal entry or not, it can nonetheless allow diagonalizability when no fewer than 4 non-diagonal entries exist. Therefore, the result holds[7].

From Theorem 6.2.4, according to traversing method, there are 78 kinds of 3 order sign patterns that do not allow diagonalizability. Moreover, there exist 3^{16} kinds of 4 order sign patterns, so it is very difficult to verify if a sign pattern allows diagonalizability again by traversing method. Therefore, we will consider if a 4 order sign pattern allows diagonalizability by cycles and minimum rank theory. Below is the results of 4 order sign patterns.

Theorem 6.2.5 Let $A \in Q_4$. If A allows diagonalizability, then it only can be during the following cases:

(1) $A = 0$

(2) $c(A) \geqslant 3$

(3) $c(A) = 1$ and $mr(A) = 1$

(4) $c(A) = 2$

(4a) At most either one or two non-zero diagonal entries are present in sign patterns and $mr(A) = 2$. However, any 2-order principal minors are non-zero by either 2-cycle (if any) or two diagonal entries.

(4b) Two non-zero diagonal entries are present in sign patterns, $mr(A) \leqslant 2$, and any 2-order principal minors are non-zero by 2-cycle or two diagonal entries. They are the following matrices (or are permutationally similar to):

$$\begin{bmatrix} a_{11} & a_{12} & a_{13} & a_{14} \\ a_{21} & a_{22} & a_{23} & a_{24} \\ 0 & 0 & 0 & 0 \\ 0 & 0 & 0 & 0 \end{bmatrix}, \begin{bmatrix} 0 & 0 & a_{13} & a_{14} \\ 0 & 0 & a_{23} & a_{24} \\ 0 & 0 & a_{33} & a_{34} \\ 0 & 0 & a_{43} & a_{44} \end{bmatrix}, \begin{bmatrix} 0 & 0 & 0 & 0 \\ 0 & a_{22} & a_{23} & a_{24} \\ 0 & a_{32} & a_{33} & a_{34} \\ 0 & 0 & 0 & 0 \end{bmatrix},$$

$$\begin{bmatrix} 0 & a_{12} & a_{13} & 0 \\ 0 & a_{22} & a_{23} & 0 \\ 0 & a_{32} & a_{33} & 0 \\ 0 & 0 & 0 & 0 \end{bmatrix}, \begin{bmatrix} 0 & a_{12} & a_{13} & a_{14} \\ 0 & a_{22} & a_{23} & a_{24} \\ 0 & a_{32} & a_{33} & a_{34} \\ 0 & 0 & 0 & 0 \end{bmatrix}$$

($a_{14} \neq 0$, at most a zero entry exists in a_{12} and a_{13} or also at most a zero entry exists in a_{24} and a_{34}) and their transpose matrices.

Proof Let $A \in Q_4$. Thus, $0 \leqslant c(A) \leqslant 4$. We then discuss if A allows diagonalizability by $c(A)$.

(1) $c(A) = 0$. If $A = 0$, then A allows diagonalizability. If $A \neq 0$, in terms of Theorem 2.2.2.3., we can not find out an non-singular principal submatrix, then A does not allow diagonalizability.

(2) $c(A) \geqslant 3$. By Theorem 2.1.5, obviously A allows diagonalizability.

(3) $c(A) = 1$. Thus, the cycle is a self-loop, which means that a non-zero diagonal entry exists in A. If $mr(A) = 1$, then a $B \in Q(A)$ with rank 1 exists,

which has a 1×1 non-singular principal submatrix, and $\boldsymbol{A}$ allows diagonalizability. If $mr(\boldsymbol{A})>1$, then we can not find out an non-singular principal submatrix of Theorem 2.2.2.3., and $\boldsymbol{A}$ does not allow diagonalizability.

(4) $c(\boldsymbol{A})=2$. We will discuss the problem according to two cases.

(i) At most one or two non-zero diagonal entries are present in sign patterns, and $mr(\boldsymbol{A})=2$. However, any 2-order principal minors are non-zero by either 2-cycle (if any) or two diagonal entries. In this case, we can identify a 2×2 non-singular principal submatrix for any $\boldsymbol{B}\in Q(\boldsymbol{A})$ with rank 2. Thus, A allows diagonalizability. If $mr(\boldsymbol{A})>2$, then $\boldsymbol{A}$ does not allow diagonalizability.

(ii) Two non-zero diagonal entries are present in sign patterns, $mr(\boldsymbol{A})\leqslant2$, and any 2-order principal submatrices are non-zero by either 2-cycle or two diagonal entries. Thus, the minimum rank of 2-order principal submatrices by 2-cycle (if any) or two diagonal entries is 1. If A allows diagonalizability, then we must identify a $k\times k$ non-singular principal submatrix for $\boldsymbol{B}\in Q(Q)$ with rank $k(k=1$ or $2)$. Because permutational similarity does not change diagonalizability, in Terms Lemma 6.2.1, we only need to consider the following 3 kinds of Frobenius normal forms and their transpose matrices:

$$\boldsymbol{A}_1=\begin{bmatrix} a_{11} & a_{12} & a_{13} & a_{14} \\ a_{21} & a_{22} & a_{23} & a_{24} \\ 0 & 0 & 0 & a_{34} \\ 0 & 0 & 0 & 0 \end{bmatrix};\ \boldsymbol{A}_2=\begin{bmatrix} 0 & a_{12} & a_{13} & a_{14} \\ 0 & a_{22} & a_{23} & a_{24} \\ 0 & a_{32} & a_{33} & a_{34} \\ 0 & 0 & 0 & 0 \end{bmatrix};\ \boldsymbol{A}_3=\begin{bmatrix} 0 & a_{12} & a_{13} & a_{13} \\ 0 & 0 & a_{23} & a_{24} \\ 0 & 0 & a_{33} & a_{34} \\ 0 & 0 & a_{43} & a_{44} \end{bmatrix}.$$

For $\boldsymbol{A}_1$, if it allows diagonalizability, then $a_{34}=0$. Then, we can identify a 2×2 non-singular principal submatrix for $\boldsymbol{B}\in Q(\boldsymbol{A})$ with rank 2, and thus, $\boldsymbol{A}_1$ allows diagonalizability.

Similarly, if $\boldsymbol{A}_3$ allows diagonalizability, then $a_{12}=0$.

For $\boldsymbol{A}_2$, the following results are available:

If $a_{12}=a_{13}=a_{14}$ or $a_{14}=a_{24}=a_{34}=0$, then $\boldsymbol{A}_2$ allows diagonalizability.

When $a_{14}\neq0$ and $a_{12}=a_{13}=0$ or $a_{24}=a_{34}=0$, $\boldsymbol{A}_2$ does not allow diagonalizability. When only one of a_{12} and a_{13} is zero or only one of a_{24} and a_{34} is zero, then we can solve for a 2×2 non-singular principal submatrix for $\boldsymbol{B}\in Q(\boldsymbol{A})$ with rank 2. Thus, $\boldsymbol{A}_2$ allows diagonalizability.

When $a_{12}a_{13}a_{14}\neq0$ and $a_{14}a_{24}a_{34}\neq0$, if $mr(\boldsymbol{A}_2)=1$, then $\boldsymbol{A}_2$ allows diagonalizability. If $mr(\boldsymbol{A}_2) = 2$, then we can solve for a 2×2 non-singular principal submatrix for $\boldsymbol{B} \in Q(\boldsymbol{A})$ with rank 2. Thus, $\boldsymbol{A}_2$ allows diagonalizability.

Obviously their transpose matrices also have the similar results. Therefore, the theorem holds. [7]

Remark 6. 2. 6 From Theorem 6. 2. 5, we can find that 4 order sign patterns have many forms, and whether a 4 order sign pattern allows diagonalizability is also complex. Especially, we consider if $\boldsymbol{A}_2$ allows diagonalizability, we need consider it by sign of entries of matrices. Moreover, for this kind of sign patterns, at present we still have not effective combinational method to judge whether a sign pattern allows diagonalizability. This will be a future work for us. Moreover, this is also an important significance for us to analyse if a sign pattern allows diagonalizability from low order sign patterns.

At present the necessary and sufficient conditions from combinational opinion is still an open problem that sign pattern allows diagonalizability. Therefore, in this section we obtained some results of 1-4 order sign patterns that allow diagonalizability. We want to find its rules and finally to solve it. From proof of Theorem 6. 2. 5, if we want to solve this open problem, except for minimum rank and composite cycles, we also need to introduce more tools.

6. 3 Recent advance on sign patterns with special graphs that allow diagonalizability

In this part, we will introduce some Indian researcher Das's results about allowing diagonalizability, In 2022, Das S [7] further considered sign patterns that allow diagonalizability whose graphs are star or path, and presented a sufficient condition of sign pattern matrices that allow diagonalizability whose graphs are trees. In the following content we will introduce them.

Definition 6. 3. 1 [7] Let $\boldsymbol{A}$ be a tree sign pattern matrix of order $\boldsymbol{A}$. An index i is said to be an essential index of $\boldsymbol{A}$ if there is a maximum matching M in $G(\boldsymbol{A})$ such that $i \notin V(M)$ and every principal submatrix of $\boldsymbol{A}$ whose index set contains $V(M) \cup \{i\}$ requires singularity. An index i is said to be non-essential if it is not an essential index.

Theorem 6. 3. 2 [7] Let $\boldsymbol{A}$ be a sign pattern matrix such that $G(\boldsymbol{A})$ is a tree and A requires singularity. If there are no directed paths between the essential indices of the principal submatrices corresponding to two distinct strong components of $D(A)$, or if all possible directed paths between the essential indices of the principal submatrices corresponding to two distinct strong components of $D(A)$ contain non-essential indices from at least one of those strong components, then there exists a diagonalizable matrix $B \in Q(A)$ with $rank(B) = MR(A)$.

Theorem 6.3.3[7] Let A be a sign pattern matrix such that $G(A)$ is a path. Then $(+, -)$ allows diagonalizability if and only if A allows non-singularity, or A requires singularity and there are no directed paths between the essential indices of the principal submatrices corresponding to any two distinct strong components of $D(A)$.

6.4 Research direction of sign patterns that allows diagonalizability in the future

Finally, we want to give out some suggestions about research direction of sign patterns that allow diagonalizability in the future. We think that we can do the following works in the future:

(1) Irreducible sign patterns that allow diagonalizability. We hope that we can get thoroughly some sufficient and necessary conditions for an irreducible sign pattern to allow diagonalizability by adding some extra conditions. By this way, further reducible sign patterns that allow diagonalizability can be solved out.

(2) Higher order sign patterns that allow diagonalizability. Through higher order sign patterns that allow diagonalizability, we hope to find out more laws about this problem. Ultimately, we hope this problem can be solved out.

(3) Sign patterns in Frobenius normal form. This problem still deserves further to be studied. We hope that better results can be found in this field.

(4) Some open problems of sign patterns that allow diagonalizability.

Notation

$\mathbf{R}$	the real numbers
C	the complex numbers
F	usually a field (usually $\mathbf{R}$ or C)
$\mathbf{N}^*$	the set of natural numbers
$M_n(\mathbf{R})$	$n \times n$ matrices with entries from $\mathbf{R}$
Q_n	the set of all $n \times n$ sign patterns
$Q(\boldsymbol{A})$	the qualitative class of $\boldsymbol{A}$
#	ambiguous number (the result of adding + with −)
$D(\boldsymbol{A})$	either the signed digraph (see page 91) or $n - mr(\boldsymbol{A})$ (see page 4)
γ	a cycle
$c(\boldsymbol{A})$	the maximum cycle length of $\boldsymbol{A}$
$MR(\boldsymbol{A})$	the maximal rank of $\boldsymbol{A}$
$mr(\boldsymbol{A})$	the minimal rank of $\boldsymbol{A}$
$\boldsymbol{A}^{\mathrm{T}}$ or $\boldsymbol{A}'$	the transpose matrix of $\boldsymbol{A}$
$\boldsymbol{A}^{-1}$	the inverse of non-singular $\boldsymbol{A}$
$\boldsymbol{B}^*$	conjugate transpose matrix of $\boldsymbol{B}$
J	all the entries of J are +
$\boldsymbol{I}_n$	the identity sign pattern of order n
$\det\boldsymbol{A}$	determinant of $\boldsymbol{A}$
$p_{\boldsymbol{B}}(\mathrm{t})$	the characteristic polynomial of $\boldsymbol{B}$
$E_i(\boldsymbol{B})$	the sum of all i-by-i principal minors of $\boldsymbol{B}$
$S_k(\boldsymbol{B})$	the kth elementary symmetric function of the eigenvalues of $\boldsymbol{B}$
$f'(x)$	the derivative of $f(x)$

$rank(\boldsymbol{B})$	the rank of $\boldsymbol{B}$
$\boldsymbol{A}[X,Y]$	the submatrix of $\boldsymbol{A}$ with row index set X and column index set Y
$\boldsymbol{A}[X]$	the principal submatrix of $\mathbf{A}$ whose row index set (as well as the column index set) is X
$\lvert S \rvert$	the index number of index set S
T^c	complement of the set T
$z(\boldsymbol{B})$	the algebraic multiplicities of 0 as an eigenvalue of real matrix $\boldsymbol{B}$
$g(\boldsymbol{B})$	geometric multiplicities of 0 as an eigenvalue of real matrix $\boldsymbol{B}$
$z(\boldsymbol{A})$	the minimum algebraic multiplicity of 0 as an eigenvalue of a matrix in $Q(\boldsymbol{A})$
$g(\boldsymbol{A})$	the minimum geometric multiplicity of 0 as an eigenvalue of a matrix in $Q(\boldsymbol{A})$
$Z(\boldsymbol{A})$	the maximum algebraic multiplicity of 0 as an eigenvalue of a matrix in $Q(\boldsymbol{A})$
$G(\boldsymbol{A})$	either the maximum geometric multiplicity of 0 as an eigenvalue of a matrix in $Q(\boldsymbol{A})$, see page 72, or the underlying undirected graph$D(\boldsymbol{A})$, see page 91
$\text{sgn}(\boldsymbol{B}) = \boldsymbol{A}$	means $\boldsymbol{B} \in Q(\boldsymbol{A})$
$\text{Res}(f(x), g(x))$	the resultant of two polynomials $f(x)$ and $g(x)$
$\text{discr}(f(x))$	the discriminant of a polynomial $f(x)$
$\gamma_1 \subset \gamma_2$	there is a composite cycle β_2 such that $\gamma_2 = \gamma_1\beta_2$

Index

References

[1] ALON N, SPENCER J H. The probabilistic method [M]. 2nd ed. New York: Wiley, 2000.

[2] ARAV M, HALL F, LI Z, et al. Minimum ranks of sign patterns via sign vectors and duality [J]. Electronic Journal of Linear Algebra, 2015, 30: 360 - 371.

[3] BRUALDI R A, RYSER H J. Combinatorial matrix theory [M]. Cambridge: Cambridge University Press, 1991.

[4] BRUALDI R A, SHADER B L. Matrices of sign-solvable linear systems [M]. Cambridge: Cambridge University Press, 1995.

[5] CHOI M D, HUANG Z J, LI C K, et al. Every invertible matrix is diagonally equivalent to a matrix with distinct eigenvalues [J]. Linear Algebra and Its Applications, 2012, 436: 3773 - 3776.

[6] CULOS G J, OLESK DD, DRIESSCHE P. Using sign patterns to detect the possibility of periodicity in biological systems [J]. Journal of Mathematical Biology, 2016, 72: 1281 - 1300.

[7] DAS S. Sign patterns associated with some graphs that allow or require diagonalizability [J]. Electronic Journal of Linear Algebra, 2022, 38: 131 - 159.

[8] DELSARTE P, KAMP Y. Low rank matrices with a given sign pattern [J]. Siam Journal on Discrete Mathematics. 1989, 2: 51 - 63.

[9] ESCHENBACH C A, HALL F J, JOHNSON C R, et al. The graphs of the unambiguous entries in the product of two (+, -) sign pattern matrices [J]. Linear Algebra and Its Applications, 1997, 260: 95 - 118.

[10] ESCHENBACH C A, JOHNSON C. Sign patterns that require repeated eigenvalues [J]. Linear Algebra and Its Applications, 1993, 190: 169 - 179.

[11] FANG W, GAO W, GAO Y, et al. Minimum ranks of sign patterns and zero-nonzero patterns and point-hyperplane configurations [J]. Linear Algebra and Its Applications, 2018, 558: 44 - 62.

[12] FENG X L, LI Z, HUANG T Z. Is every non-singular matrix diagonally equivalent to a matrix with all distinct eigenvalues? [J]. Linear Algebra and Its Applications, 2012, 436: 120 - 125.

[13] FENG X L, HUANG T Z, LI Z, et al. Sign patterns that allow diagonalizability revisited [J]. Linear and Multilinear Algebra, 2013, 61: 1223 - 1233.

[14] FENG X L. Sign pattern matrices that allow k-potent and diagonalizability [J]. Journal of Leshan Normal University, 2017, 8: 1 - 4.

[15] FENG X L, GAO W, HALL F J, et al. Minimum rank conditions for sign patterns that allow diagonalizability [J]. Discrete Mathematics, 2020, 343: 111798.

[16] FENG X L, LI Z. Some results on sign patterns that allow diagonalizability [J]. Journal of Mathematical Research with Applications, 2022, 42: 111 - 120.

[17] FORSTER J. A linear lower bound on the unbounded error probabilistic communication complexity [J]. Journal of Computer and System Sciences, 2002, 65: 612 - 625.

[18] HALL F J, LI Z, WANG D. Symmetric sign pattern matrices that require unique inertia [J], Linear Algebra and Its Applications. 2001, 338: 153 - 169.

[19] HALL F J, LI Z. Sign pattern matrices, in handbook of linear algebra, chap. 42, l. hogben, ed. [M]. Boca Raton: CRC Press, 2014.

[20] HALL F J, LI Z. Sign Pattern Matrices, chapter 33 in handbook of linear algebra [M]. Boca Raton: CRC Press, 2007.

[21] HORN R A, JOHNSON C R. Matrix analysis [M]. Cambridge: Cambridge University Press, 1985.

[22] HORN R A, LOPATIN A K. The moment and Gram matrices, distinct eigenvalues and zeroes, and rational criteria for diagonalizability [J]. Linear Algebra and Its Applications, 1999, 299: 153 - 163.

[23] HORN R A, JOHNSON C R. Matrix analysis, 2nd ed [M]. Cambridge: Cambridge University Press, 2013.

[24] JACOBSON N. Basic algebrai [M]. San Francisco: W. H. Freeman, 1974.

[25] JOHNSON C R. Lecture notes [R]. Atlanta: Combinatorial Matrix Theory Workshop Held at Georgia State University, 1991.

[26] KIM P J. On the 4 by 4 irreducible sign pattern matrices that require four distinct eigenvalues [D]. Thesis: Georgia State University, 2011.

[27] KONSTANTINOV M M, PETKOV P H, GU D W, et al. Nonlocal sensitivity analysis of the eigensystem of a matrix with distinct eigenvalues [J]. Numerical Functional Analysis and Optimization, 1997, 18: 367 - 382.

[28] LEE S G, PARK J W. Sign idempotent sign pattern matrices that allow idempotence [J]. Linear Algebra and Its Applications, 2015, 487: 232 - 241

[29] LI Z, HARRIS L. Sign patterns that require all distinct eigenvalues [J]. JP Journal of Algebra, Number Theory and Applications, 2002, 2: 161 - 179.

[30] LI Z, HALL F J, ZHANG F. Sign patterns of non-negative normal matrices [J]. Linear Algebra and Its Applications, 1997, 254: 335 - 354.

[31] LI Z, GAO Y, ARAV M, GONG F, et al. Sign patterns with minimum rank 2 and upper bounds on minimum ranks [J]. Linear and Multilinear Algebra, 2013, 61: 895 - 908.

[32] LING Y L, WANG B C. First-and second-order eigensensitivity of matrices with distinct eigenvalues [J]. International Journal of Systems Science, 1988, 19: 1053 - 1067.

[33] MISHRA B. Algorithmic algebra [M]. New York: Springer Verlag, 1993.

[34] RAZBOROV AA, SHERSTOV A A. The sign-rank of AC^0 [J]. Siam Journal on Computing, 2010, 39: 1833 - 1855.

[35] SAMUELSON P A. Foundations of economic analysis [M]. Cambridge: Harvard University Press, 1947.

[36] SHAO Y L, GAO Y B. Sign patterns that allow diagonalizability [J].

Linear Algebra and Its Applications, 2003, 359: 113 - 119.

[37] SHAO Y L, GAO Y B. On Eschenbach-Johnson conjecture about diagonalizability [J]. Piournal of Algebra, Number Theory and Applications, 2006, (6): 573 - 583.

[38] STUART J, ESCHENBACH C, KIRKLAND S. Irreducible sign k-potent sign pattern matrices [J]. Linear Algebra and Its Applications, 1999, 294: 85 - 92.

[39] STURMFELS B. Solving systems of polynomial equations [M]. Providence: American Mathematical Society, 2002.

[40] THOMASSEN C. Sign-nonsingular matrices and even cycles in directed graphs [J]. Linear Algebra and Its Applications, 1986, 75: 27 - 41.

Acknowledge

This book is intended to provide the fundamental material for researchers of sign pattern matrices that allow diagonalizability. It also include my several publications about sign pattern matrix.

It is with great pleasure that I express my appreciation to all those who have expressed support, enthusiasm and encouragement in this adventure. I am forever indebted to my families and close friends for their patience, understanding and support.

Firstly, I am grateful to Prof. Tingzhu Huang (University of Electronic Science and Technology of China), who is my doctoral supervisors, and led me into this research field. Secondly, I am grateful to Prof. Zhongshan Li (Georgia State University), who provided me much of the guidance and help in this field. Finally, I am grateful to Prof. Mingli Luo (Leshan Normal University), who provided many suggestions and help for the writing and publication of this book. I also thank our college leader Prof. Xiaobing Qu, Secretary Liqiang Ma etc. who provided many supports for the publiction of this book.